OUR UNIVERSE VIA DREXLER DARK MATTER

Drexler Dark Matter Created and Explains Dark Energy, Top-Down Cosmology, Inflation, Accelerating Cosmos, Stars, Galaxies, Cosmic Web

JEROME DREXLER

Universal-Publishers
Boca Raton

OUR UNIVERSE VIA DREXLER DARK MATTER:
Drexler Dark Matter Created and Explains Dark Energy,
Top-Down Cosmology, Inflation, Accelerating Cosmos, Stars,
Galaxies, Cosmic Web

Universal-Publishers
Boca Raton, Florida • USA
2009

ISBN-10: 1-59942-887-3
ISBN-13: 978-1-59942-887-1

www.universal-publishers.com

Library of Congress Cataloging-in-Publication Data

Drexler, Jerome.
  Our universe via Drexler dark matter : Drexler dark matter
created and explains dark energy, top-down cosmology, inflation,
accelerating cosmos, stars, galaxies, cosmic web / Jerome Drexler.
      p. cm.
  Includes bibliographical references and index.
  ISBN-13: 978-1-59942-887-1 (pbk. : alk. paper)
  ISBN-10: 1-59942-887-3 (pbk. : alk. paper)
  1. Dark matter (Astronomy) 2. Cosmology. I. Title.
  QB791.3.D74 2009
  523.1'126--dc22

                                        2009041595

This book is dedicated to Sylvia, my wife, best friend,
and lifelong partner and to my father, Max Drexler,
who taught me about the cosmology of
Nicolaus Copernicus and the wonders of astronomy.

"Scientific advancement is not evolutionary,
but rather is a series of peaceful interludes punctuated
by intellectually violent revolutions."

~Thomas S. Kuhn ~

"Leave the beaten track occasionally and dive into the woods.
Every time you do so you will find something you have never seen
before. Follow it up, explore all around it, and before you know
it, you will have something to think about to occupy your mind.
All really big discoveries are the result of thought."

~ Alexander Graham Bell ~

"We are to admit no more causes of natural things than such
are both true and sufficient to explain their appearances."

~ Issac Newton's version of Occam's razor ~

"Make everything as simple as possible, but not simpler."

~ Albert Einstein ~

"It is dangerous to be right in matters on which
the established authorities are wrong."

~ Voltaire ~

FORWARD

Mainstream cosmology is in a crisis.

These three paragraphs, which were part of the introduction to the Paris symposium, "Invisible Universe — Toward a New Cosmological Paradigm", June 29 - July 10, 2009, explain it:

"Cosmology has arrived at a crossroads. According to the best data available, from large ground-based telescopes and space observatories, almost 95 percent of the universe irretrievably escapes observational detection."

"This missing part of the cosmos is constituted for 25 percent by a mysterious form of dark matter and 70 percent, by a dark energy whose nature is even more exotic and unknown! But what are exactly these new physical entities?"

"In an attempt to answer this complex and profound question, more than 400 experts will gather in Paris to evaluate the situation, and draw future perspectives. The basic principles of physics appear sometimes to be put into question. Modern cosmology is perhaps at the beginnings of a major renewal, similar to those once made by Galileo and Einstein."

Drexler's Cosmology offers a solution.

Amazon.com Book Reviews for his 2008 Book:
Discovering Postmodern Cosmology

"This third book in a series by Drexler shows how his thesis, that dark matter is composed of charged ultra high energy relativistic protons, is capable of solving up to 25 previously unresolved mysteries of the Cosmos. Older cold dark matter concepts, now generally discredited, relied on too few observations and have required additional hypotheses to account for each new experimental finding. In significant contrast: each new data set gathered subsequent to Drexler's first publication of his thesis has appeared to reinforce his concepts without the need for adaptation. Most recently the publication by astronomers at the University of Chicago titled "Reopening the window on charged dark matter" which occurred 6 months after Drexler's third book first became available, lends considerable additional support to the thesis that dark matter is composed of charged particles. While this book is sure to prove controversial amongst conservative astrophysicists, I would encourage the reader to keep an open mind. Remember there was a time when conventional wisdom had it that the sun revolved around a flat earth!"

"Discovering challenges to conventional wisdom is always interesting and especially so when the challenge is as compelling and well-reasoned as in this book. Jerome Drexler presents a plausible theory as to the composition of the dark matter that represents a high percentage of the mass of the universe but whose makeup mystifies cosmologists. Drexler posits that this dark matter consists of relativistic protons, which he believes are capable of forming galaxies, dark matter, the cosmic web, and newborn stars. Throughout the book, he repeatedly demonstrates how conventional cosmology is frequently at odds with actual astronomical observations and even with the laws of physics. Drexler's Postmodern Cosmology model presents a coherent theory that solves a number of cosmological "mysteries", including the nature of the Big Bang."

"A must read for anyone interested in straightforward, logical solutions to some of cosmology's most significant unsolved problems. This book highlights all of Drexler's work to date, particularly his relativistic-baryon dark matter hypothesis; it also establishes his primacy on the concept of a Relativistic Big Bang that satisfies the second law of thermodynamics. It is now incumbent on the scientific community to accept the plausibility of Drexler's theories, and to find additional supporting observational evidence (e.g., the UV signatures of Lyman-alpha blobs), or to refute them based on such evidence."

"Drexler presents a conceptually coherent and logically appealing model for the mechanics underlying the large scale structure of the universe. As his theory departs dramatically from the current Standard Cosmological Model, it will certainly attract vociferous criticism. Yet ongoing reports of newly observed and measured astronomical phenomena seem to be, more often than not, congruent with Drexler's Halo theory for dark matter and dark energy. Until his hypotheses are definitively falsified, the fact that some of the phenomena required by Drexler's theory have not yet been observed is no more troubling than the current non-observation of WIMPs, neutralinos, or MACHOs. Altogether, this book presents an intriguing rationalization for the many mysteries currently unexplained by the Standard Cosmological Model."

PREFACE

Physicists say that no one understands the nature of dark matter; even though it was discovered 75 years ago and represents about 23% of the mass of the universe.

Physicists say that no one understands the nature of dark energy or what causes the accelerating expansion of the universe; even though they were discovered 11 years ago and dark energy represents about 70% of the mass-energy of the universe.

Physicists say that no one knows where ultra-high-energy cosmic-ray protons, that bombard Earth's atmosphere every day, derive their very high energies; even though they were discovered over 90 years ago.

Physicists say that no one understands the nature of the post-big-bang inflationary epoch phenomenon of very rapid expansion of the universe, commonly called Cosmic Inflation; even though it was discovered 28 years ago.

I am pleased to report that through seven years of intensive cosmology research and good fortune, I have been able to discover a unified theory of astrophysical cosmology that plausibly solves these four long-unsolved mysteries, as well as more than a dozen other cosmic mysteries.

This book describes this unified astrophysical cosmology theory and how it is utilized to solve these four famous mysteries as well as a number of others, in thirty-one chapters. This new cosmology paradigm also can provide insights into and solutions to newly discovered cosmic mysteries within days or weeks following the publication of the related scientific papers.

Most of the excitement and success during the past 50 years in the field of astrophysical cosmology has stemmed from the discovery of *new astronomical mysteries* rather than from solving then existing mysteries. There does not seem to be enough properly trained cosmologists and astrophysicists to slow the rising tide of unsolved cosmic mysteries, enigmas, anomalies, discrepancies, and conundrums.

Meanwhile, astronomers are building more and more advanced telescopic systems by utilizing space platforms, employing adaptive optics, and by combining images derived from photons of different wavelengths. With more and more cosmic mysteries being discovered and the slow progress in solving them, cosmologists and astrophysicists must re-train themselves to understand and also to utilize the postmodern unified astrophysical cosmology model and to maximize the knowledge derived from the astronomical data. These are the three principal objectives of this book.

I previously documented six years of my dark matter/dark energy research, its timeline, its interaction with mainstream cosmology, and the overwhelming evidence that relativistic-proton dark matter represents the principal constituent of the dark matter of the universe in the six publications listed here:

(1) Scientific Web site dated Dec. 8, 2008, entitled, "Discovering Dark Matter Cosmology" at: http://www.jeromedrexler.org/.

(2) Paperback book, March 1, 2008, *Discovering Postmodern Cosmology: Discoveries in Dark Matter, Cosmic Web, Big Bang, Inflation, Cosmic Rays, Dark Energy, Accelerating Cosmos.*

(3) Scientific paper, physics/0702132, Feb. 15 2007, "A Relativistic-Proton Dark Matter Would Be Evidence the Big Bang Probably Satisfied the Second Law of Thermodynamics".

(4) Paperback book, May 22, 2006, *Comprehending and Decoding the Cosmos: Discovering Solutions to Over a Dozen Cosmic Mysteries by Utilizing Dark Matter Relationism, Cosmology, and Astrophysics.*

(5) Scientific paper, astro-ph/0504512, April 22, 2005, "Identifying Dark Matter through the Constraints Imposed by Fourteen Astronomically Based 'Cosmic Constituents'".

(6) Paperback book, Dec. 15, 2003, *How Dark Matter Created Dark Energy and the Sun: An Astrophysics Detective Story.*

This book is different from all other modern cosmology books in several ways. It introduces a cosmologic universe, which is orderly, logical, and systematic. It teaches and explains by illustrating how a variety of cosmic mysteries have been solved. It raises the status of dark matter in the universe by illuminating its roles as the principal source of energy, the principal source of matter in the form of hydrogen and helium, and the principal source of cosmic relationships with the principal cosmic phenomena and cosmic constituents of the universe. This book simplifies the universe as the book of Nicolaus Copernicus simplified the solar system in 1543.

CONTENTS

Our Universe via Drexler Dark Matter

Drexler Dark Matter Created & Explains Dark Energy, Top-Down Cosmology, Inflation, Accelerating Cosmos, Stars, Galaxies, Cosmic Web

INTRODUCTION

This book, *Our Universe via Drexler Dark Matter* was so named because by using the Drexler relativistic-proton dark matter model, along with the laws of physics, and known astronomical data, at least 18 cosmic constituents or cosmic phenomena of the universe were determined and can be explained plausibly.

The author takes a step further with the subtitle and states that Drexler dark matter, with its protons, helium nuclei, and its enormous kinetic energy actually created dark energy, top-down cosmology, cosmic inflation, the accelerating cosmos, the stars, galaxies, and the Cosmic Web. Dark matter seems to be playing the role of stem cells in a developing human embryo that can transform themselves into all of the specialized embryonic tissues needed to create a fully functioning human body.

The Drexler dark matter model and its related dark energy model and related cosmic-ray proton model were announced on December 15, 2003 in the author's first book, "How Dark

Matter Created Dark Energy and the Sun. Six years and many astronomical observations later the three 2003 cosmologic models remain unchanged. Meanwhile Drexler has published two more books and two scientific papers that are fully compatible with the three 2003 models. Furthermore, the same three 2003 models are compatible with top-down cosmology, the Cosmic Web, cosmic inflation, star formation, galaxy formation and other cosmic constituents and cosmic phenomena.

The author believes that this book's 31 chapters provide the evidence necessary to support both the main title and subtitle. This also means that at least 18 cosmic constituents or cosmic phenomena have dark matter as a "common parent." This common-parent linkage implies that the Drexler dark matter cosmology model probably represents a unified astrophysical cosmology for our universe and makes our universe appear to be relatively orderly and logical.

In contrast, for the past 25 years mainstream physicists have believed that based upon astronomical observations and theory, the dark matter of the universe cannot be made of protons or neutrons or anything that was once made of protons or neutrons. According to them, calculations of particle synthesis during the big bang indicate that such proton and neutron based particles were simply too few in number to make up the estimated mass of dark matter in the universe.

Based upon these 1984 calculations [1], scientists have searched for a dark matter that contains neither protons nor neutrons, called non-baryonic matter. They have not been successful. They even overlooked the relativistic-proton dark matter; the key solution to their "missing mass" problem, that Drexler discovered in early 2002 and announced in 2003.

Physicists still have not found any evidence of the putative non-baryonic Cold Dark Matter WIMPs (weakly interacting massive particles) for which they have been searching for 25 years. Moreover, the physicists' WIMP-based cosmology has been unable to explain any of the following 18 cosmic constituents or cosmic phenomena known to exist in the universe, even though Drexler relativistic-proton dark matter provides plausible cosmologic explanations for each and every one of them.

A list follows of the 18 cosmic constituents or cosmic phenomena that are the focus of this book. The list includes the chapter numbers that are relevant to each subject.

1. The accelerating expansion of the universe *(see Chapters 9, 15, 19, 21, F)*.

2. Dark Energy *(see Chapters 9, 15,19, 21, F)*.

3. A dark matter that can exist in the form of spheroidal halos around spiral disk galaxies and also in the form of long large slightly curved filaments that form the Cosmic Web *(see Chapters 1, 2, 3, 10, 15, 16, 17, 18, B, C, D, E, J).*

4. Source of ultra-high-energy cosmic-ray protons that bombard Earth's atmosphere *(see Chapters 10, 15, C, G, H).*

5. How Cosmic Inflation started then stopped during the big bang period *(see Chapters 21, H).*

6. Why most large galaxies formed without galaxy mergers *(see Chapters 6, 11, 14).*

7. The causes for the early rapid growth of massive galaxies *(see Chapters 6, 11, 14).*

8. The causes for the stunted mass growth of galaxy clusters *(see Chapters 9, 13, 22).*

9. How the first stars formed without availability of hydrogen molecules or dust *(see Chapter 20).*

10. The basis for the formation of the Lyman Alpha blobs *(see Chapter 12).*

11. The limitation of the diameter of galaxy superclusters to 430 million light years *(see Chapter10).*

12. Top-Down theory of galaxy formation *(see Chapters 6, 11, 14, 17, D).*

13. Causes of ultraviolet (UV), EUV, or soft X-ray photon emission from dark matter *(see Chapters 8, 12, 13, 22, F).*

14. NASA discovers loud synchrotron-emission microwave noise *(see Chapters 5, 7).*

15. The nature of the Cosmic Web *(see Chapters 13, J).*

16. Roles of muons in cosmology *(see Chapter12, 13, 20).*

17. New view of the nature of the big bang *(see Chapter C).*

18. Relativistic protons orbiting galaxies may be evading GZK cosmic-ray cutoff effect *(see Chapter G).*

This group of 18 sources of cosmologic evidence provides overwhelming support for the Drexler relativistic-proton dark matter. In contrast, mankind has waited 25 years for non-baryonic Cold Dark Matter WIMPs to prove that they exist and represent about 83% of the mass of the universe. They haven't done either. The 1984 non-baryonic WIMP theory[1] appears to have reached a cosmological dead end.

To demonstrate the validity and significance of his relativistic-proton dark matter model, Drexler used it to solve two dozen cosmic mysteries and published the results in *Comprehending and Decoding the Cosmos*, in May 2006 and *Discovering Postmodern Cosmology*, in March 2008.

Drexler, a Bell Labs-trained scientist/inventor, has completed seven years researching dark matter/dark energy cosmology and accumulating evidence supporting his unified astrophysical cosmology theory. This book is Drexler's capstone vehicle, along with his three earlier books and two scientific papers, to complete the launch of his unified astrophysical cosmology. Hopefully, the reader will feel that the book's cosmological evidence probably ensures that the Drexler dark matter cosmology paradigm will be adopted.

Some of the early history of dark matter research can be found in the first four pages of Chapter C. An 20-page glossary and 80 references are provided.

CHAPTER 1

University of Chicago's CHAMPs Dark Matter Boosts Drexler's Dark Matter over WIMPs

September 18, 2008 — On Sept. 2, 2008, the University of Chicago's Department of Astronomy and Astrophysics, published a scientific paper online entitled, "Reopening the Window On Charged Dark Matter" [2].

The paper's dark matter, in the form of electrically charged massive particles (CHAMPs), boosts Bell Labs-trained scientist Jerome Drexler's five-year-old relativistic-proton dark matter model and undermines the 24-year-old putative Cold Dark Matter theory of uncharged weakly interacting massive particles (WIMPs), also known as neutralinos.

Drexler's dark matter theory launched Drexler's postmodern cosmology theory that simultaneously answers fundamental questions about dark matter, the big bang, cosmic inflation, the accelerating cosmos, ultra-high-energy cosmic rays, and the Cosmic Web.

The last sentence of the abstract of the University of Chicago paper (arXiv:0809.0436 v1) gives clues as to the paper's significance. It reads, "Further, we find that charged massive particles [CHAMPs] may simultaneously solve several long-standing astrophysical problems, including the under abundance of dwarf galaxies, the shallow [mass] density profiles in the cores of the LSB [low surface brightness] galaxies, the absence of cooling flows in the cores of galaxy clusters, and several others."

Solving long-standing astrophysical problems was also the goal of Drexler's three books and two online scientific papers. He uses relativistic-proton dark matter that simultaneously solves over 15 astrophysical problems, mysteries, dilemmas, or conundrums. Note that Drexler's dark matter particles are the only known real-world manifestation of CHAMPs. His three books were written as a trilogy with the first published December 2003, the second May 2006, and the third March 2008.

The University of Chicago paper makes a good prequel to Drexler's trilogy since it provides a compelling introduction to the December 2003 book. For science enthusiasts, a *NewScientist.com* news article on Sept. 9 entitled, "Is dark matter a wimp or a champ?" [3] could function as a prequel to Drexler's trilogy. (Note that a dark matter WIMP is a cold uncharged weakly interacting massive particle, a dark matter CHAMP is a charged massive particle and a relativistic dark matter particle is a proton or helium nucleus that becomes as massive as a CHAMP by moving at relativistic velocities.)

Drexler utilizes the evidence provided in his three books, his two scientific papers, and the University of Chicago paper to stake his claim to the discovery of the precise identity of the long-sought dark matter of the universe, which was first publicly disclosed in his December 15, 2003 book.

These five publications cover the precise nature of dark matter, the evidence supporting that conclusion, and the relationships that dark matter has with dark energy, the accelerating expansion of the universe, cosmic rays, the big bang, cosmic inflation, and the Cosmic Web. These cosmic relationships are keys to precisely identifying the dark matter of the cosmos. Since dark matter represents about 83 percent of the mass of the universe, any dark matter candidate that does not have relationships with most of these six cosmic phenomena should be treated with suspicion.

These five Drexler publications also disclose dark matter's surprising and significant roles and functions in creating the spiral galaxies, stars, starburst galaxies and ultra-high-energy cosmic rays.

(1) Book, March 1, 2008, *Discovering Postmodern Cosmology: Discoveries in Dark Matter, Cosmic Web, Big Bang, Inflation, Cosmic Rays, Dark Energy, Accelerating Cosmos.*

(2) Scientific paper, physics/0702132, Feb. 15, 2007, "A Relativistic-Proton Dark Matter Would Be Evidence the Big Bang Probably Satisfied the Second Law of Thermodynamics".

(3)   Book, May 22, 2006, *Comprehending and Decoding the Cosmos: Discovering Solutions to Over a Dozen Cosmic Mysteries by Utilizing Dark Matter Relationism, Cosmology, and Astrophysics.*

(4)   Scientific paper, astro-ph/0504512, April 22, 2005, "Identifying Dark Matter through the Constraints Imposed by Fourteen Astronomically Based 'Cosmic Constituents'".

(5)   Book, Dec. 15, 2003, *How Dark Matter Created Dark Energy and the Sun: An Astrophysics Detective Story.*

CHAPTER 2

Doubts Cast on Cold Dark Matter by Cambridge, Cardiff U, CEA Saclay, NYU, Russian Academy of Sciences, UC San Diego

November 10, 2008 — A recent article on the formation of galaxies, in the journal *Nature*, has undermined the credibility of the Cold Dark Matter (CDM) theory of WIMPs at the expense of UC Santa Cruz, CDM's principal supporter and originator in 1984. The Physicsworld article about the Nature paper is entitled, "Galaxy survey casts doubt on cold dark matter" [4].

The UC Santa Cruz (UCSC) central doctrine for CDM has been that small galaxies form first and larger galaxies are formed through mergers of smaller galaxies. This is called hierarchal galaxy formation, a central principle of the UCSC Cold Dark Matter theory. This should lead to a non-simple galaxy formation process based upon a half dozen independent variables representing various galaxy parameters.

The October 23, 2008 *Nature* article, authored by Professor Michael J. Disney of UK's Cardiff University and five associates, is entitled, "Galaxies appear simpler than expected" [5]. It turns out that through a statistical analysis of the radio and optical data from 200 galaxies, five of the six independent variables actually are dependent on some single unknown independent variable. The last sentence of the abstract makes a key statement, "Such a degree of organization [of galaxies] appears to be at odds with hierarchical galaxy formation, a central tenet of the cold dark matter model in cosmology." This is a strong and potentially prize-winning challenge to UC Santa Cruz's Cold Dark Matter thesis.

More from Professor Disney's abstract: "Here we report that a sample of galaxies that were first detected through their neutral hydrogen radio-frequency emission, and are thus free from optical selection effects shows five independent correlations among six independent observables, despite having a wide range of properties. This implies that the structure of these galaxies must be controlled by a single parameter, although we cannot identify this parameter from our data set. Such a degree of organization [of galaxies]... appears to be at odds with hierarchical galaxy formation, a central tenet of the cold dark matter model in cosmology."

In addition to the six universities/research establishments mentioned in the above title, doubts have also been cast on UC Santa Cruz's Cold Dark Matter by the University of Chicago, Harvard, and by Jerome Drexler when he was a Research Professor in physics in 2005 at the New Jersey

Institute of Technology, and afterward when he authored two more astro-cosmology books when based in Silicon Valley.

The following is a sampling of articles and papers casting doubt on the UC Santa Cruz theory of uncharged, proton-free, and hydrogen-free Cold Dark Matter with its weakly interacting massive particles (WIMPs), which have not been detected after two decades of searching:

October 2008 paper, "Galaxies appear simpler than expected" [5] by Cardiff University's Professor Michael J. (Mike) Disney, et al, published in *Nature*. The PhysicsWorld article about this paper is entitled, "Galaxy survey casts doubt on cold dark matter."

September 2008 paper, "Reopening the Window on Charged Dark Matter," [2] by University of Chicago Professor Edward F. (Rocky) Kolb, et al, published as astro-ph arXiv:0809.0436 v1: "Further, we find that charged massive particles [CHAMPs] may simultaneously solve several long-standing astrophysical problems, including the under abundance of dwarf galaxies, the shallow [mass] density profiles in the cores of the LSB [low surface brightness] galaxies..."

September-October 2007 paper "Modern Cosmology: Science or Folk Tale" [6] by UK's Cardiff University Professor Michael J. Disney, published in American Scientist magazine, Volume 95. "This situation [Lambda-Cold Dark Matter] is very far from healthy."

September 2007 newswire article, "NASA Data Raises Doubts of Existence of Cold Dark Matter in Galaxy Clusters" [7] by Jerome Drexler, published in *Discovering Postmodern Cosmology,* Chapter 24.

May 2007 paper, "Missing Mass in Collisional Debris from Galaxies" [8] by Dr. F. Bournaud, et al, (CEA Saclay, France) published in Science 25 May 2007, Vol.316 no.5828, p.1166-1169. "It more likely indicates that a substantial amount of dark matter resides within the disks of spiral galaxies. The most natural candidate is molecular hydrogen in some hard to trace form."

A May 2007 newswire, "'Ring of Dark Matter' Uncovered from Anomalies-Discrepancies"[9] by Jerome Drexler, is also published as Chapter D. The top-down theory of galaxy formation used in Drexler's postmodern cosmology solves the anomalies/discrepancies dilemma, whereas the bottom-up theory of galaxy formation of UCSC's Cold Dark Matter appears to have created the dilemma.

March 2007 paper, "The Observed properties of Dark Matter on small spatial scales" [10] by Cambridge Professor Gerard Gilmore, et al, published as arXiv:astro-ph/0703308v1. "Galaxy formation models inside the Lambda CDM [Cold Dark Matter] paradigm however have considerable difficulties matching observations on small scales. The well-known 'satellite problem' is an example, as is the 'cores vs. cusps' debate."

February 2007 paper, "A Relativistic-Proton Dark Matter Would Be Evidence the Big Bang Probably Satisfied the Second Law of Thermodynamics" [11] by Jerome Drexler published as arXiv physics/0702132v1 and in Chapter C. It is unlikely the Second Law could be satisfied by a big bang producing high-entropy Cold Dark Matter WIMPs.

October 2006 paper, "A New Force in the Dark Sector" [12] by NYU Professor G.R.Farrar, et al, published as arXiv: astro-ph/0610298v1. "The number of superclusters observed in SDSS [Sloan Digital Sky Survey] data appears to be an order of magnitude larger than predicted by Lambda-Cold Dark Matter simulations."

June 2006 paper, "Cold Dark Matter Cosmology Conflicts with Fluid Mechanics and Observations" [13] by UC San Diego Professor Carl H. Gibson, published online in the arXiv June 2006 as astro-ph/0606073 and also in the Journal of Applied Fluid Mechanics.

June 2006 news: Russia announced it will launch an ultraviolet astronomical observatory in 2010 having a 1.7 meter main mirror. The project manager is Boris Shustov, Professor of Physics and Mathematics and head of the Institute of Astronomy at the Russian Academy of Sciences.

The release quotes him, "One should particularly emphasize the observatory's role in detecting the so-called dark matter of the Universe and unlocking its secrets because such dark matter can only be seen by large ultraviolet telescopes" [14]. The proponents of Cold Dark Matter make no claim for UV emission. Drexler's relativistic protons do emit UV.

April 2005 paper, "Identifying Dark Matter through the Constraints Imposed by Fourteen Astronomically Based 'Cosmic Constituents'" [15] by then NJIT Research Professor Jerome Drexler published as arXiv astro-ph/0504512v1. The paper's analysis of the possible relationships of 14 cosmic constituents with dark matter makes a strong case for relativistic-proton dark matter over Cold Dark Matter WIMPs.

March 1990 paper, "Charged dark matter" [16] by Nobel Laureate Harvard Professor Sheldon L. Glashow, et al, published in Nucl. Phys. B, Part. Phys., Vol. 333, No.1. From a 1989 interview: "People have been excluding the possibility of charged dark matter for no good reason and limiting themselves to neutral particles," says physicist Sheldon L. Glashow of Harvard University. "If you don't know what dark matter is, it would seem wise to be open-minded." "Glashow and his collaborators propose that dark matter consists of stable, very massive, electrically charged elementary particles left over from the Big Bang."

Drexler utilizes the overwhelming evidence provided in his three books, his two scientific papers, the papers of Harvard's Prof. Glashow and Chicago's Prof. Kolb, and those researchers listed above casting doubt on the existence of Cold Dark Matter WIMPS, to stake his claim to the discovery of the precise identity and physics of the universe's dark matter, which he first publicly disclosed in his December 15, 2003 book.

CHAPTER 3

Relativistic-Proton Dark Matter Explains Cosmic Web, Accelerating Cosmos, Inflation, UHECRs, Big Bang

December 8, 2008 — Scientists at UC Santa Cruz claim that about 85 percent of the mass of the universe is Cold Dark Matter (CDM), comprised of cold, uncharged, weakly interacting massive particles (WIMPs) that contain no nuclei or atoms of hydrogen, helium, oxygen, nitrogen or any other atoms/nuclei. Such extraordinary claims require extraordinary evidence to be convincing, yet UC Santa Cruz has not provided any astronomical evidence after 24 years of research.

The new astronomy-oriented 'Dark Matter Cosmology' website at http://www.jeromedrexler.org, featuring Jerome Drexler's recent astro-cosmology books, *Discovering Postmodern Cosmology* and *Comprehending and Decoding the Cosmos,* provides overwhelming evidence of the precise identity of the universe's dark matter, which is unrelated to UC Santa Cruz's Cold Dark Matter WIMPs. The term "astro-cosmology" used here means astronomy-oriented dark matter cosmology.

Furthermore, the new Web site provides substantial evidence supporting plausible explanations for the following six famous mysteries of cosmology: the nature of dark matter, the accelerating expanding cosmos, the effect of the Second Law of Thermodynamics on the big bang, the hyper expansion of cosmic inflation immediately after the big bang, the source of ultra-high-energy cosmic rays (UHECRs), and the nature of the Cosmic Web.

The search for the true dark matter of the universe has suffered from a psychological fog that has prevented scientists from determining dark matter's precise make-up. This fog has precluded scientists from recognizing the true nature of dark matter even when they encountered its overwhelming evidence.

This obscuring and confusing fog was created by the dilemma of a theoretical Cold Dark Matter that still remains undetected after 24 years of searching, but nevertheless has been accepted by the mentally exhausted mainstream astrophysicists and cosmologists as a comfortable default solution. This dilemma and its fog have seriously hampered dark matter astronomical research efforts for more than a decade.

Encouragingly, during the past 30 months innovative, courageous, and dynamic astronomers and physicists have published their doubts about the existence of Cold Dark

Matter at seven well-known universities or research establishments. They are the UK's Cambridge and Cardiff University, France's CEA Saclay, NYU, the Russian Academy of Sciences, UC San Diego and the University of Chicago.

Two other universities were even earlier. In April 2005, a paper, "Identifying Dark Matter through the Constraints Imposed by Fourteen Astronomically Based 'Cosmic Constituents'", was published by then physics research professor Jerome Drexler of New Jersey Institute of Technology, as arXiv astro-ph/0504512 v 1. The paper's analysis of the possible relationships of 14 cosmic constituents with dark matter makes a strong case for Drexler's relativistic-proton dark matter over Cold Dark Matter WIMPs.

In March 1990, a paper, "Charged dark matter" [16] by physics Nobel Laureate Harvard Professor Sheldon L. Glashow, et al, was published in Nucl. Phys. B, Part. Phys.,Vol. 333, No.1. In a 1989 interview, related to his research, he is quoted: "People have been excluding the possibility of charged dark matter for no good reason and limiting themselves to neutral particles," says physicist Sheldon L. Glashow of Harvard University.

"If you don't know what dark matter is, it would seem wise to be open-minded." "Glashow and his collaborators propose that dark matter consists of stable, very massive, electrically charged elementary particles left over from the Big Bang." [Drexler's relativistic-proton dark matter matches this.]

Drexler dedicates his astronomy-oriented "Dark Matter Cosmology" Web site to the world's astronomers and to NASA, the National Science Foundation (NSF), and the US Department of Energy (US DOE).

CHAPTER 4

Cold Dark Matter (WIMP) Doubters
Are Publishing More Scientific Papers

December 19, 2008 — This month, two dark-matter researchers at University of California, Irvine raised additional doubts about the existence of Cold Dark Matter's weakly interacting massive particles (WIMPs). They published a paper in Physical Review Letters entitled, "Dark-Matter Particles without Weak-Scale Masses or Weak Interactions" [17].

The first sentence of their abstract reads, "We propose that dark matter is composed of particles that naturally have the correct thermal relic density, but have neither weak-scale masses nor weak interactions."

Thus, the researchers at UC Irvine have joined those at UC San Diego, University of Chicago, Cambridge, Cardiff University, Harvard, NYU, NJIT, CEA Saclay, Russian Academy of Sciences, and NASA who have published papers doubting the existence of UC Santa Cruz's theoretical dark matter WIMPs. This renaissance in cosmology is documented in Jerome Drexler's Web site at http://www.jeromedrexler.org and in his books.

The UC Santa Cruz (UCSC) central doctrine for WIMP-based Cold Dark Matter (CDM) has been that small galaxies form first and large galaxies are formed through mergers of the small galaxies. This is called hierarchal galaxy formation, a central principle of the UCSC Cold Dark Matter theory of weakly interacting massive particles (WIMPs), which leads to a non-simple galaxy formation process based upon a half dozen independent variables representing various galaxy parameters.

The History of Cold Dark Matter Doubters

A recent astronomically-based scientific paper on the formation of galaxies, in the journal *Nature*, has undermined the credibility of the Cold Dark Matter (CDM) theory of WIMPs at the expense of UC Santa Cruz, the principal supporter of Cold Dark Matter and its originator in 1984. The Physicsworld article about the *Nature* paper is entitled, "Galaxy survey casts doubt on cold dark matter".

The *Nature* paper [5], authored by Professor Michael J. Disney of the United Kingdom's Cardiff University and five associates, is entitled, "Galaxies appear simpler than expected." It turns out that through a statistical analysis of the radio and optical data from 200 galaxies, five of the six independent variables actually are dependent on some single unknown independent variable. The last sentence of the abstract makes a key statement, "Such a degree of organization [of galaxies] appears to be at odds with hierarchical galaxy formation, a central tenet of the cold dark matter model in cosmology."

The following is a sampling of other articles and papers casting doubt on the UC Santa Cruz theory of uncharged, proton-free, hydrogen-free, and atom-free Cold Dark Matter with its weakly interacting massive particles (WIMPs), which have not been detected after two decades of scientific searching.

September 2008 scientific paper [2], "Reopening the Window on Charged Dark Matter," by University of Chicago Professor Edward F. (Rocky) Kolb, et al, published as astro-ph arXiv:0809.0436 v1. "Further, we find that charged massive particles [CHAMPs] may simultaneously solve several long-standing astrophysical problems, including the underabundance of dwarf galaxies, the shallow [mass] density profiles in the cores of the LSB [low surface brightness] galaxies..."

September-October 2007 scientific paper [6], "Modern Cosmology: Science or Folk Tale" by UK's Cardiff University Professor Michael J. Disney, published in American Scientist magazine, Volume 95. This situation [Lambda-Cold Dark Matter] is very far from healthy."

September 2007 Newswire article [7], "NASA Data Raises Doubts of Existence of Cold Dark Matter in Galaxy

Clusters," by Jerome Drexler, published in *Discovering Postmodern Cosmology,* Chapter 24.

May 2007 paper [8], "Missing Mass in Collisional Debris from Galaxies" by Dr. F. Bournaud, et al, (CEA Saclay, France) published in Science 25 May 2007, Vol.316 no.5828, p.1166-1169. "[I]t more likely indicates that a substantial amount of dark matter resides within the disks of spiral galaxies. The most natural candidate is molecular hydrogen in some hard to trace form."

May 2007 newswire article [9], "'Ring of Dark Matter' Uncovered from Anomalies-Discrepancies" by Jerome Drexler, also published as Chapter D. The top-down theory of galaxy formation used in Drexler's postmodern cosmology solves the anomalies/discrepancies dilemma, whereas the hierarchal bottom-up theory of galaxy formation of UCSC's Cold Dark Matter theory appears to have created the dilemma.

March 2007 paper [10], "The Observed properties of Dark Matter on small spatial scales" by Cambridge Professor Gerard Gilmore, et al, published as arXiv:astro-ph/0703308v1. "Galaxy formation models inside the Lambda CDM [Cold Dark Matter] paradigm however have considerable difficulties matching observations on small scales. The well-known 'satellite problem' is an example, as is the 'cores vs. cusps' debate."

February 2007 paper [11], "A Relativistic-Proton Dark Matter Would Be Evidence the Big Bang Probably Satisfied the Second Law of Thermodynamics" by Jerome Drexler, published as arXiv physics/0702132v1 and in Chapter C. It is unlikely the Second Law was satisfied by a big bang producing high-entropy Cold Dark Matter WIMPs.

October 2006 paper [12], "A New Force in the Dark Sector?" by NYU Professor G.R.Farrar, et al, published as arXiv: astro-ph/0610298v1. "The number of superclusters observed in SDSS [Sloan Digital Sky Survey] data appears to be an order of magnitude larger than predicted by Lambda-Cold Dark Matter simulations."

June 2006 scientific paper [13], "Cold Dark Matter Cosmology Conflicts with Fluid Mechanics and Observations" by UC San Diego Professor Carl H. Gibson, published online in the arXiv June 2006 as astro-ph/0606073 and also in the Journal of Applied Fluid Mechanics.

June 2006 news [14]: Russia announced it will launch an ultraviolet astronomical observatory in 2010 having a 1.7 meter main mirror. The project manager is Boris Shustov, Professor of Physics and Mathematics and head of the Institute of Astronomy at the Russian Academy of Sciences. The release quotes him, "One should particularly emphasize the observatory's role in detecting the so-called dark matter of the Universe and unlocking its secrets because such dark matter can only be seen by large ultraviolet telescopes." The proponents of Cold Dark Matter make no claim for UV emission. Drexler' relativistic protons do emit UV.

Two of the earliest doubters of UC Santa Cruz's Cold Dark Matter WIMP theory, who also authored scientific papers on the subject, were Professor Sheldon L. Glashow at Harvard in March 1990 and Research Professor Jerome Drexler in April 2005, then with the New Jersey Institute of Technology (NJIT).

April 2005 scientific paper [15], "Identifying Dark Matter through the Constraints Imposed by Fourteen Astronomically Based 'Cosmic Constituents' " by NJIT Research Professor Jerome Drexler published as arXiv astro-ph/0504512v1. The paper's analysis of the possible relationships of 14 cosmic constituents with dark matter makes a strong case for relativistic-proton dark matter over Cold Dark Matter WIMPs.

March 1990 scientific paper [16], "Charged dark matter" by Nobel Laureate Harvard Professor Sheldon L. Glashow, et al, published in Nucl. Phys. B, Part. Phys.,Vol. 333, No. 1. From a 1989 interview: "People have been excluding the possibility of charged dark matter for no good reason and limiting themselves to neutral particles," says physicist Sheldon L. Glashow of Harvard University. "If you don't know what dark matter is, it would seem wise to be open-minded." "Glashow and his collaborators propose that dark matter consists of stable, very massive, electrically charged elementary particles left over from the Big Bang."

CHAPTER 5

NASA Discovers Loud Synchrotron-Emission Radio Noise; Source May Be Relativistic-Proton Dark Matter

January 15, 2009 — Bell-Labs trained scientist Jerome Drexler has authored a trilogy of astro-cosmology books and two scientific papers during the past five years supporting his 2003 claim that the long-sought dark matter of the universe is comprised of multitudinous galaxy-orbiting relativistic protons generating a high level of synchrotron emission throughout the universe as they cross magnetic field lines.

Such galaxy-orbiting relativistic protons would generate synchrotron emission over a broad band spectrum encompassing wavelengths from microwaves to x-rays. Doubting cosmologists have argued that such synchrotron emission from a relativistic-proton dark matter would have been detected years ago.

This month, NASA space balloon researchers reported their surprising discovery of a mysterious extra-loud radio noise that permeates the universe. It appears to be an extragalactic, synchrotron-emission-based radio noise, six times more intense than anyone predicted. This discovery is based upon measurements in microwave frequency bands of 3, 8, 10, 30, and 90 gigahertz, which peaked in the 3 and 8 GHz detectors. Considering the enormous strength and distribution of the detected radio noise, it conceivably could be generated by Drexler's relativistic-proton dark matter.

One of the four January 5, 2009 papers of the researchers is entitled, "ARCADE 2 Observations of Galactic Radio Emission"[18] and can be found at (http://arxiv.org/abs/0901.0562v1). ScienceDaily headlined its related article, "NASA Space Balloon Mission Tunes in to Cosmic Radio Mystery" [21]. SpaceDaily's headline reads, "Loud noise permeates cosmos, NASA says". Science News' heading is, "Tuned in to New Noise from the Cosmos", Sky & Telescope's headline is, "New Cosmic Background Radiation Found", and that of Fox News is "Mystery Roar Detected From Faraway Space".

The NASA researchers discovered the surprisingly strong, uniformly distributed, radio noise power at a level estimated at six times higher than the combined radio emission from all known radio sources in the universe. The spectrum of such radio noise is consistent with that produced by radio galaxies via charged particles spiraling across magnetic field lines. This phenomenon is called   synchrotron emission, which is electromagnetic radiation that is emitted from electrically-charged particles moving at relativistic velocities across transverse magnetic field lines, which accelerate the particles orthogonally.

The loud microwave radio noise is not accompanied by infrared thermal emission as in the case of well-known radio galaxies. But it is accompanied by bremsstrahlung emission radio noise, which is caused by the rapid deceleration of electrically-charged relativistic particles during collisions. The researchers are convinced the noise source does not match any known pattern from sources in the Milky Way and is not from distant galaxies or from decaying particles of exotic dark matter. The above features and characteristics are compatible with Drexler's relativistic-proton dark matter, which thus becomes a logical and plausible candidate for the microwave radio-noise source.

Confidence in the high-power microwave-noise (radio-noise) discovery was enhanced by a retrospective analysis of several other radio-noise studies in the 1980s and 1990s that hint at the unexpected loud radio noise and also by the fact that the loud radio noise was observed in a part of the microwave spectrum that had not been well studied.

The researchers also provided the following information: The noise-detection instrument, launched in July 2006 from NASA's Columbia Scientific Balloon Facility in Palestine, Texas, flew to an altitude of 120,000 feet. The researchers base their findings on 2.5 hours of data gathered during a flight of seven radio receivers called ARCADE (Absolute Radiometer for Cosmology, Astrophysics, and Diffuse Emission). ARCADE's radio receivers, which were cooled to a temperature just 2.7 degrees above absolute zero for the balloon flight are the first detectors capable of identifying the mysterious radio-noise signals. *(This study of their research is continued in Chapter 7.)*

CHAPTER 6

Most Large Galaxies Formed Without Mergers, Say Jerome Drexler, Michael J. Disney, and Avishai Dekel

January 28, 2009 — Leading-edge researchers in the field of galaxy formation have published their findings that most large galaxies have formed and developed without the involvement of galaxy mergers.

Avishai Dekel of the Hebrew University published his scientific paper [19] supporting the top-down theory of galaxy formation this month, Michael J. Disney of Cardiff University of Wales published [5] his November 2008 and Jerome Drexler published his April 2005 [15]. (Later supporting this theory was the University of Hawaii in March 2009 [47,48] and the Liverpool John Moores University in April 2009 [28].)

The UC Santa Cruz (UCSC) central doctrine for Cold Dark Matter has been that small galaxies form first and larger galaxies are formed through mergers of smaller galaxies.

This is called hierarchal galaxy formation, a central principle of the UCSC Cold Dark Matter WIMP theory. Such a hierarchal galaxy merging procedure would probably lead to a complex galaxy formation process based upon a number of independent variables representing various parameters of the merging galaxies.

The October 23, 2008 *Nature* article, authored by Professor Michael J. Disney of UK's Cardiff University and five associates, is entitled, "Galaxies appear simpler than expected" [5].

It turns out that through a statistical analysis of the radio and optical data from 200 galaxies, five of the six "independent" variables actually are dependent on some single unknown independent variable. The last sentence of the abstract makes a key statement, "Such a degree of organization [of galaxies] appears to be at odds with hierarchical galaxy formation, a central tenet of the cold dark matter model in cosmology."

More from Professor Disney's abstract: "Here we report that a sample of galaxies that were first detected through their neutral hydrogen radio-frequency emission, and are thus free from optical selection effects shows five independent correlations among six independent observables, despite having a wide range of properties. This implies that the structure of these galaxies must be controlled by a single parameter, although we cannot identify this parameter from our data set. Such a degree of organization appears to be at odds with hierarchical galaxy formation, a central tenet of the cold dark matter model in cosmology."

Professor Avishai Dekel of the Hebrew University of Jerusalem, with nine associates, comes to the same general conclusion as Disney via a different set of data and different arguments in a *Nature* article dated January 22, 2009. It is entitled, "Cold streams in early massive hot haloes as the main mode of galaxy formation" [19]. A January 25 news release from The Hebrew University begins with the following title and first three paragraphs:

"New understanding of the origin of galaxies advanced by Hebrew U astrophysicists"

"A new theory as to how galaxies formed in the Universe billions of years ago has been formulated by Hebrew University of Jerusalem cosmologists. The theory takes issue with the prevailing view on how the galaxies came to exist.

"The new theory, motivated by advanced astronomical observations and based on state-of-the-art computer simulations, maintains that the galaxies primarily formed as a result of intensive cosmic streams of cold gas (mostly hydrogen) and not, as the current theory contends, due primarily to galactic mergers. The researchers show that these mergers had only limited influence on the cosmological makeup of the universe as we know it.

"The galaxies are the building blocks of the Universe. Every galaxy is embedded in a spherical halo made of dark matter that cannot be seen but is detected through its massive gravitational attraction. The exact nature of this matter is still unknown."

There are currently two schools of thought on galaxy formation. There is the bottom-up theory, supported by the vast majority of the world's universities, which states that small galaxies form first and larger galaxies are formed through mergers of the small galaxies.

The principal subject of this article is the top-down theory of galaxy formation, that Drexler, Disney, and Dekel support, which generally states that galaxies form and grow via some source of hydrogen not involving galaxy mergers.

Bell-Labs trained scientist Jerome Drexler has authored a trilogy of astro-cosmology books and two scientific papers during the past five years supporting his 2003 claim that the long-sought dark matter of the universe is comprised of multitudinous galaxy-orbiting relativistic protons, which orbit groups of galaxies as well as individual galaxies. He contends that many relativistic cosmic-ray protons that enter Earth's atmosphere are dislodged straggler protons that had been dark-matter-halo protons orbiting the Milky Way.

Drexler explains his top-down theory of galaxy formation in his May 22, 2006 book, entitled *Comprehending and Decoding the Cosmos,* in Chapters 19, 21, 31, 36, 40 and 41. His March 1, 2008 book, entitled *Discovering Postmodern Cosmology,* discusses the top-down theory of galaxy formation in Chapters 9 and 19.

Drexler's well-proven relativistic-proton dark matter theory permits a more complete top-down theory of galaxy formation than that provided by others. Drexler's May 2006 book's definition of his top-down theory is "that long, large dark matter filaments form galaxy clusters where the dark matter filaments intersect/collide and then galaxies form from the remnants of these collisions." Drexler's March 2008 book's definition of the top-down theory is the same except for the addition of the then new words "of the cosmic web" between the words "filaments" and "form".

Thus in Drexler's 2006 galaxy-formation theory the evolving star-forming galaxies are fed with streams of warm-hot protons directly from the relativistic-proton dark matter itself rather than from a posited separate source of protons or hydrogen. Note that Drexler's star-forming system has Occam razor simplicity.

CHAPTER 7

Proton Swarms from Drexler's Dark Matter May Fill Universe; Create Microwave Noise Detected by NASA

February 12, 2009 – (Continued from Chapter 5.) A loud cosmic microwave noise recently discovered by a NASA balloon may be generated by synchrotron emission via multitudinous swarms of energy-exhausted protons zipping through magnetic fields throughout the universe. The origin of these swarms of "tired" relativistic protons could be relativistic-proton dark matter. Let us look into this.

Nine years ago the famous CERN (Organisation Europeenne pour la Recherche Nucleaire) research establishment on the Franco-Swiss border published a graph in the CERN Courier (Vol. 35, No.10) entitled, "Cosmic-Ray Energy Distribution at the Earth"[20]. It depicts the density of cosmic-ray protons at energy levels from 10(9) to 10(20) electron-volts (eV). Although the graph may look innocuous, it may carry a key to unlocking the 75-year old mystery of the precise nature of the dark matter of the universe.

The significance of the CERN cosmic-ray graph to the subject of dark matter became increasingly apparent to Bell Labs-trained scientist Jerome Drexler as he was reading about the January 5, 2009, news that NASA space balloon researchers had discovered a mysterious extra-loud radio noise that permeates the universe. The static-like noise signal, coming from all directions, appears to be an extragalactic, synchrotron-emission-based microwave noise, six times more intense than anyone predicted. This discovery is based upon noise measurements in microwave frequency bands at 3, 8, 10, 30, and 90 gigahertz, which peaked in NASA's detectors at 3 and 8 GHz. One of the four January 5, 2009 papers of the researchers is entitled, "ARCADE 2 Observations of Galactic Radio Emission"[18] and can be found at http://arxiv.org/abs/0901.0562v1.

ScienceDaily headlined its related article, "NASA Space Balloon Mission Tunes in to Cosmic Radio Mystery" [21]. *SpaceDaily's* headline reads, "Loud noise permeates cosmos, NASA says". *Science News'* heading is, "Tuned in to New Noise from the Cosmos"; *Sky & Telescope's* headline is, "New Cosmic Background Radiation Found"; and that of *Fox News* is "Mystery Roar Detected From Faraway Space".

The NASA balloon researchers discovered the surprisingly strong, isotropic, extragalactic distributed, microwave noise power at a level estimated at six times higher than the combined microwave emission from all known microwave-radio sources in the universe. The spectrum of such microwave noise is consistent with that produced, by what are known as radio galaxies, via electrons spiraling in a magnetic field, which thereby emit microwave-radio noise through synchrotron emission. Synchrotron emission is electromagnetic radiation that is emitted from electrically-charged particles (protons or electrons) moving at relativistic velocities across transverse magnetic field lines, which accelerate the particles orthogonally.

The loud microwave noise is not accompanied by infrared thermal emission as in the case of well-known radio galaxies. But appears to be accompanied by bremsstrahlung emission radio noise, which is caused by the rapid deceleration of electrically charged relativistic particles during collisions. Apparently, the researchers believe the noise source does not match any known pattern from sources in the Milky Way or its halo and is not from some distant galaxies or from decaying particles of exotic dark matter.

Based upon the extragalactic distribution and the specific power-peaks of synchrotron microwave noise at the two lowest frequencies, the noise conceivably could have been generated by enormous swarms of energy-reduced relativistic-proton stragglers from Drexler's relativistic-proton dark matter. The logic is as follows.

We know from the NASA balloon data that the peak synchrotron-emission microwave noise power occurs with detectors at the microwave frequencies of 3 and 8 GHz and we know the extragalactic magnetic field is about 10(-9) gauss, which are sufficient in conjunction with available astrophysics formulas to calculate the corresponding proton energies, which turn out to about 1.1 and 1.8 x 10(9) eV. (See *High Energy Cosmic Rays,* Section 2.3.1, by Todor Stanev.)

The CERN graph shows that compared to a one proton per square meter per second flux for 10(11) eV protons, for proton energies at 1.1 and 1.8 x10(9) eV the proton flux is close to 10,000 times higher! Thus for protons with energies below 10(10) eV the total quantity of relativistic protons appears to represent a very significant fraction of all the relativistic protons zipping through the universe. (The CERN graph can be found on page 24 of Drexler's 2003 book, page 243 of his 2006 book, and page 203 of his 2008 book and in Appendix II of this book.)

Note that the synchrotron emission from a proton with energy of only 10(9) eV is so low that its subsequent kinetic energy decline, from continuing synchrotron emission, would be negligible. The end result could be a universe filled with swarms of lowest-energy relativistic protons each generating low level, synchrotron-emission noise in or near the microwave range of 3 to 8 gigahertz that was detected by NASA. But the overall grand total of microwave synchrotron emission power from such multitudinous proton swarms would be so large that they might represent a new form of cosmic microwave background radiation.

Drexler's relativistic-proton dark matter theory posits that there were not many low-energy protons immediately after the big bang; such low-energy protons evolved via synchrotron-emission energy losses and collisions and accumulated in swarms over billions of years. From the CERN graph data, it is possible to posit that swarms of energy-reduced straggler protons from relativistic-proton dark matter are a logical and plausible source of the loud microwave noise detected by the NASA balloon.

Note that the March 1990 paper by Nobel Laureate Harvard Prof. Sheldon L. Glashow, et al., entitled, "Charged Dark Matter," the September 2008 paper of University of Chicago's Prof. Rocky Kolb, et al., entitled, "Reopening the Window on Charged Dark Matter", the nature of the 1999 CERN cosmic-ray energy graph, and the loud microwave noise detected by the NASA balloon are all compatible with the relativistic-proton dark matter model.

CHAPTER 8

Two Key Signatures of Drexler's Dark Matter Are Found in the Leo Ring: Is This Meaningful?

March 3, 2009 — Researchers, using the Galaxy Evolution Explorer (GALEX) orbiting space telescope, reported in the journal *Nature* on February 19 that far-ultraviolet (FUV) photon emissions emanating from the Leo ring exhibit peaks of emission from regions of concentrations of neutral atomic hydrogen.

The Leo ring, which is about 35 million light-years away in the constellation Leo, orbits a pair of elliptical galaxies (M105 and NGC 3384). The *Nature* paper is entitled, "Massive star formation within the Leo 'primordial' ring"[22].

Four articles about the *Nature* paper include such statements as, "Galaxy Mix: No Dark Matter Required"[23] from Science News, "Strangely, the new galaxies appear to lack dark matter," "The cloud, known as the Leo Ring, appears to lack the dark matter," and "Experts thought that dark matter was a prerequisite for the birth of galaxies." Apparently, the researchers concluded that the long-accepted Cold Dark Matter is not present in the Leo ring.

Both the observed FUV photon emission wavelengths and the FUV emission links to neutral atomic hydrogen in the

Leo ring are key signatures of relativistic-proton dark matter, another form of dark matter, discovered in 2002 by Jerome Drexler. Does this imply that Drexler's dark matter may be present in the Leo ring, but not Cold Dark Matter? Since this would be a significant discovery, let us explore this possibility.

Bell Labs-trained scientist Jerome Drexler, previously published (online on April 16, 2007 and in Chapter 17 of his March 1, 2008 book) his estimates for the peak power synchrotron-emission photon wavelengths for relativistic-proton dark matter orbiting a group of galaxies and also for such dark matter orbiting a single galaxy. The geometric mean of these two peak-power wavelengths turns out to be very compatible with the observed FUV photon emission wavelengths reported in *Nature* for the Leo ring, which orbits two galaxies.

It is not surprising that neutral atomic hydrogen would be found where relativistic-protons have been orbiting galaxies for billions of years while colliding with dust, nuclei, and photons. However, claiming emission of FUV photons from the dark matter does call for a logical and plausible explanation. It is important to understand that synchrotron emission is electromagnetic radiation that is emitted from electrically-charged particles (protons or electrons) moving at relativistic velocities across transverse magnetic field lines that accelerate the particles orthogonally. The protons' photon emission wavelengths at peak power depend upon both proton energy and the transverse magnetic field intensity and its electromagnetic emission can be in the form of microwave, infrared, visible, ultraviolet, or x-ray photons.

Relativistic-proton dark matter orbiting a group of galaxies in a galaxy cluster is a much more likely source of Extreme UV (EUV)/soft X-ray synchrotron emission than would be relativistic-proton dark matter orbiting a single galaxy, like the Milky Way. There are three reasons for this:

(1) The wavelength of the peak synchrotron-emission photon power from a proton is inversely proportional to the square of the proton's energy,

(2) the typical energies of dark matter relativistic protons orbiting a group of galaxies in a galaxy cluster are estimated at about 30 times greater than those in a dark matter halo orbiting a single galaxy like the Milky Way, and

(3) a relativistic proton's synchrotron-emission photon power is directly proportional to the square of the proton's energy.

Accordingly, two relevant paragraphs from Chapter 17 of Drexler's March 1, 2008 book, *Discovering Postmodern Cosmology,* reads as follows: "Thus, dark matter relativistic protons orbiting groups of galaxies in the Local Group galaxy cluster should radiate synchrotron emission power about 900 times higher, at a wavelength 900 times smaller, than from protons in the Milky Way's dark matter halo.

Calculations indicate that the synchrotron emission photons from the Milky Way's dark matter halo should have a broad peak in the infrared including the wavelength of 5 microns [5000 nanometers]. Photons from protons orbiting groups of galaxies in the Local Group galaxy cluster should have an EUV/soft X-ray broad peak including the wavelength of 5.5 nanometers." (The formulas used can be found in *High Energy Cosmic Rays* [24], Section 2.3.1, by Todor Stanev.)

Thus, for dark matter relativistic protons orbiting two galaxies, such as in the case of the Leo ring, the peak synchrotron-emission power photon wavelength should be roughly equal to the geometric mean (square-root of the product) of the two wavelengths reported in the previous paragraph, or roughly 166 nm. This calculated peak-synchrotron-power wavelength is very compatible with the wavelengths actually detected by the Leo ring researchers in the FUV, which is defined as being between 200 nm and 122 nm.

Whether or not these apparently impressive results are meaningful or a mere coincidence will have to await more astronomical data analysis. For example, a spectrum analysis of the observed far-ultraviolet photon emission from the Leo ring may indicate whether it is photon emission from stars or from synchrotron emission of photons from relativistic protons.

CHAPTER 9

Drexler's Dark Matter Probably Causes the Stunted Mass-Growth of Galaxy Clusters Observed by Harvard

March 18, 2009 — A discovery was recently reported of the stunted mass-growth of galaxy clusters during the last 5 1/2 billion years, by researchers at the Harvard-Smithsonian Center for Astrophysics. This finding appears to involve the mysterious anti-gravity dark energy concept originally conceived to explain the 1998 supernova-based discovery of the accelerating expansion of the universe.

The Harvard researchers used NASA's Earth-orbiting Chandra X-ray Observatory to measure the hot gas in over 80 galaxy clusters in order to estimate the rate of mass growth for groups of galaxy clusters. Their scientific paper is entitled "Chandra Cluster Cosmology Project III: Cosmological Parameter Constraints"[25] while a Science News article about it is "Dark Energy Constantly With Us" [26].

The general consensus of the galaxy-cluster researchers and interested cosmologists is that the results are compelling and that the 1998 and recent dark energy manifestations probably represent the same or similar cosmic phenomena. The parallel successes by two different astronomical techniques have confirmed the existence of a very mysterious dark energy and give hope of further scientific progress.

Some relevant published comments by the galaxy-cluster researchers to journalists are as follows:

> "Comparing their data to models of cosmic evolution, Dr. [Alexey] Vikhlinin [of the Harvard-Smithsonian Center for Astrophysics] found that the most massive clusters are only about a fifth as plentiful today as they would be in a universe without dark energy. 'The clusters', he said, 'are still growing, but very slowly.' "

> " 'What we find is that the growth of structure [of the mass of galaxy clusters] has slowed down during the last 5 1/2 billion years, and this is unmistakably a signature of dark energy,' said Alexey Vikhlinin."

> " 'This result could be explained as arrested development of the universe,' said Alexey Vikhlinin. "This stifling of growth is the unmistakable signature of an antigravitational force that astronomers have labeled dark energy.' "

> "Dr. [Alexey] Vikhlinin lamented that there were not yet very many such theories to knock down yet, but there were sure to be more on the table soon."

> "Vikhlinin and colleagues used NASA's Chandra X-Ray Observatory (http://chandra.harvard.edu/) to measure the hot gas in 86 galaxy clusters. These groups of hundreds or thousands of galaxies are filled with 100-million-degree-gas that can best be detected with X-ray telescopes."

Is there a dark energy theory that is compatible with the supernova-based accelerating expansion of the universe observed in 1998 that also can explain the recent Harvard-Smithsonian discovery of the stunted ordinary-mass growth of galaxy clusters during the last 5 1/2 billion years? Let us try one such theory/explanation as follows:

If for some reason the mass of all the dark matter of the universe were continuously eroding and thus declining, we would not be surprised to observe the stunting of the ordinary-mass growth of galaxy clusters over time. There are two reasons for this:

Firstly, the eroding dark matter mass around each galaxy is about ten times greater than the ordinary mass of each galaxy. Secondly, the ordinary-mass growth of galaxy clusters relies upon the gravitational accretion into the clusters of nearby stars, dust, gas, and galaxies located outside the clusters, which gravitational accretion is significantly restrained by the eroding and declining dark matter mass.

Let us now consider what phenomenon could cause all the dark matter throughout the universe to be eroding and declining, as posited above. There is only one dark matter candidate whose mass is continuously eroding and declining; it is relativistic-baryon dark matter, also known as relativistic-proton dark matter, discovered by Bell Labs-trained scientist Jerome Drexler in 2002. It erodes relativistic mass through a phenomenon called synchrotron emission of photons, which comes about when relativistic protons/ baryons dart across transverse magnetic field lines in the cosmos.

Thus, both the 1998 and recent Harvard/NASA dark energy observations can be plausibly explained by means of the erosion of the dark matter mass throughout the universe via synchrotron emission of infrared, ultraviolet, and soft X-ray photons, provided that dark matter is indeed comprised primarily of relativistic-protons orbiting galaxies and groups of galaxies. There is considerable published evidence supporting the existence of relativistic-proton dark matter.

For example, the discoveries of the anti-gravity or repulsive-gravity dark energy phenomenon in 1998 and again recently, using a different astronomical technique, appear to support Drexler's dark matter/dark energy theory (see Drexler's 2003 book, *How Dark Matter Created Dark Energy And The Sun,* his 2006 book, *Comprehending And Decoding The Cosmos* and his 2008 book, *Discovering Postmodern Cosmology.*) These works disclose and explain fifteen cosmic mystery phenomena that only can be explained in a plausible manner by evoking the relativistic-proton dark matter.

Some relevant comments about this dark energy research:

"As a result, many astronomers and physicists are desperate for evidence of another explanation. Dr. [Adam] Riess said of the cosmological constant, 'The biggest thing we could learn is by ruling that out.'."

"Indeed, several theorists said the future now looked dim for alternative theories of gravity, in particular a variant from string theory, which incorporates extra dimensions and which predicts enhanced growth of structures like galaxy clusters."

" 'This is very impressive and important work,' says Charles Bennett (http://cosmos.pha.jhu.edu/bennett/), who heads NASA's Wilkinson Microwave Anisotropy Probe, a satellite that measures the big bang's afterglow. 'The results provide a crucial cross-check against the pre-existing set of cosmological results.'."

CHAPTER 10

Drexler's Dark Matter Accurately Predicts Maximum Size of 430 Million Light Years for Galaxy Superclusters

April 6, 2009 — Bell Labs-trained scientist Jerome Drexler has written three astro-cosmology books providing overwhelming evidence that the dark matter of the universe is comprised primarily of relativistic protons orbiting galaxies and groups of galaxies. Thus, a galaxy group containing ten large galaxies would be filled with swarms of relativistic protons with the lowest energy ones orbiting a single galaxy and the higher energy protons orbiting three, six, and even all ten of the galaxies.

The radius of a proton orbital path is called a Larmor Radius, which is directly proportional to the proton's energy and inversely to the orthogonal (transverse) magnetic field. Protons orbiting all ten galaxies would have the highest energy and a relativistic mass as much as one hundred million times the mass of a proton at rest. This relativistic mass of orbiting protons provides the dark matter mass to galaxies, galaxy groups and clusters, without need for new dark matter particles. Relativistic protons hold their orbits firmly via the strong electromagnetic forces that are orders of magnitude greater than gravitational tidal forces.

In contrast, the galaxies, hydrogen, helium gas, and dust in a galaxy group are bound primarily by gravitational tidal forces to the galaxy group's relativistic-proton dark matter mass, which typically is about ten times greater than the ordinary mass. Fritz Zwicky [45], then at the California Institute of Technology, discovered dark matter in 1933 when he was observing the velocities of the galaxies in the Colma galaxy supercluster and calculated there must be a large amount of unseen or missing mass (now called dark matter) to hold the high speed galaxies from being hurled into space. Colma is a relatively small galaxy supercluster.

In his analysis, Drexler inverts the 1933 supercluster-to-dark matter discovery process. He utilizes the Larmor radius of the orbital path of the highest-energy dark-matter relativistic protons that are available in quantity to calculate the maximum possible diameter-like size for galaxy superclusters. He calculates a 430 million light-year maximum size, which is very compatible with the sizes of the four largest well-known galaxy superclusters.

A galaxy supercluster typically comprises more than a thousand galaxy groups. Let us consider the general structure of galaxy groups, clusters, and superclusters, which began to be understood through a September 2004 news release from NASA and Harvard, involving the Fornax cluster, entitled, "Motions in nearby galaxy cluster reveal presence of hidden superstructure."

A key sentence in the 2004 news release reads, "Astronomers think that most of the matter in the universe is concentrated in long large filaments of dark matter and that galaxy clusters are formed where these filaments intersect." This astronomically established filamentary dark matter crisscrossing the cosmos and forming galaxy clusters where filaments collide, essentially describes Drexler's relativistic-proton dark matter. Today this "hidden superstructure" of dark matter filaments is called the Cosmic Web[33,43,77].

Each galaxy supercluster, within 1 billion light years of Earth, links together an average of 2400 galaxy groups using "long large filaments of dark matter" to form multi-strand "necklaces" of bright galaxy groups on invisible filaments of high-velocity dark matter protons. Such galaxy groups hold about 12 large galaxies. The presence of the extragalactic magnetic field of about 1 x 10(-9) gauss may indicate that each of the multi-strands of dark matter filaments probably represents a different proton velocity/energy group and that the highest energy proton group would have the largest Larmor Radius and thus would be found at the outer periphery of the galaxy supercluster structure.

Drexler used the above descriptions and concepts as a basis to calculate the "maximum possible diameter-like size of a galaxy supercluster," which is the largest contiguous cosmic structure of the universe. He started with the concept of long large filaments of relativistic-protons and their linked galaxy groups filling the universe's largest galaxy supercluster with more than 2400 galaxy groups. The highest energy protons at its outer periphery would determine the superclusters' maximum diameter-like size, which could be determined through a calculation of twice the Larmor Radius for these periphery protons.

The extragalactic magnetic field strength is already widely accepted. Therefore, the remaining task might have been to seek the highest-energy relativistic protons available to superclusters in such large quantities that they must have been derived from the big bang.

That was not necessary because Drexler already had found and interpreted such data in a cosmic-ray research paper, unrelated to dark matter.

The famous 1966 GZK cosmic-ray energy cutoff theory, for relativistic protons in open space engaged in inelastic collisions with the CMB (cosmic microwave background), was confirmed recently at 6 x 10 (19) electron-volts by the High Resolution Fly's Eye Collaboration, supported principally by the University of Utah. Their paper, published in *Physical Review Letters* in March 2008, is entitled, "First Observation of the Greisen-Zatsepin-Kuzmin [GZK] Suppression" [27]. This GZK cosmic-ray cutoff energy level should apply to all relativistic protons moving through open space in the universe, including dark-matter protons, not just cosmic-ray protons.

Using this same GZK cutoff energy level for the enormous quantities of available dark-matter protons and the extragalactic magnetic field of 1 x 10(-9) gauss, Drexler arrived at a maximum diameter-like size for galaxy superclusters at 430 million light years. He did this by calculating the Larmor Radius according to the equation on page 47 of his May 2006 book, entitled *Comprehending And Decoding The Cosmos,* and then doubling it to obtain the maximum diameter-like size for galaxy superclusters.

Is this estimate of 430 million light years for the maximum diameter-like size of galaxy superclusters logical and plausible? Are we convinced from the relativistic-proton dark matter theory that there is a maximum diameter-like size limit for galaxy superclusters? Does the 430 million light-year prediction add significant support to the relativistic-proton dark matter theory? Let us search the literature for related data about some of the largest galaxy superclusters, study what the data implies, and draw conclusions about the results and the applicability of the theory behind them. Begin with the following relevant data.

An excellent atlas of superclusters is entitled "The Universe within 1 billion Light Years—The Neighbouring Superclusters", found at http://www.atlasoftheuniverse.com/superc.html.

This atlas can be used to measure (by means of a ruler) the diameter-like contiguous sizes of galaxy superclusters against the sizes/lengths reported in the literature. The diameter-like contiguous sizes of some of the largest galaxy superclusters in the universe are as follows, listed by galaxy supercluster name and approximate diameter-like contiguous size.

Horologium-Reticulum Supercluster, 410 million light years

Sculptor Supercluster, 250 million light years

Perseus-Pisces Supercluster, 210 million light years

Shapley Supercluster, 160 million light years

Bootes Supercluster, 140 million light years

Virgo Supercluster, 110 million light years (local to Earth)

Coma Supercluster, 20 million light years (discovery of dark matter)

There is considerable published evidence supporting the existence of Drexler's relativistic-proton dark matter, upon which the 430 million light-year calculation relies. For example, the discoveries of the anti-gravity or repulsive-gravity dark energy phenomenon in 1998 and again recently by Harvard-Smithsonian, using a different astronomical technique, appear to support Drexler's published dark matter/dark energy theory. See Chapter 9 entitled, "Drexler's Dark Matter Probably Causes the Stunted Mass-Growth of Galaxy Clusters Observed by Harvard."

Also see Drexler's 2003 book, *How Dark Matter Created Dark Energy And The Sun,* his 2006 book, *Comprehending And Decoding The Cosmos* and his 2008 book, *Discovering Postmodern Cosmology.*

Furthermore, Drexler's three books provide more than fifteen cosmic-phenomena examples that justify the reliance on his relativistic-proton dark matter. These works disclose and explain these "mysterious" cosmic phenomena that only can be explained in a logical and plausible manner by evoking the relativistic-proton dark matter. The explanations include the "mysterious" source of the ultra-high-energy cosmic rays, the nature of the Cosmic Web, how the big bang satisfied the Second Law of Thermodynamics, how cosmic inflation's hyper-growth of the universe started and stopped and why the expansion of the universe is accelerating.

Are there any reader doubts about the following sentence? "This relativistic mass of orbiting protons provides the dark matter mass to galaxies, galaxy groups and clusters, without need for new dark matter particles"? If so, valuable comments on this subject can be found in Appendix I, "Some Relativity" in *Cosmic Bullets: High Energy Particles in Astrophysics* [62] by Roger Clay and Bruce Dawson.

CHAPTER 11

Nature Paper Asks, Drexler Explains How Dark Matter Caused the Early Rapid Growth of Massive Galaxies

May 4, 2009 — A paper in the journal *Nature* entitled, "Early assembly of the most massive galaxies"[28] published April 2, 2009, reports that astronomers' discovery of massive galaxies fully developed five billion years after the big bang raises serious questions about the widely accepted galaxy formation models. In these theoretical models, large galaxies grow through mergers with smaller galaxies, which is a much slower process than the rapid growth rate actually observed by astronomers. Thus, there must be some other undiscovered galaxy growth mechanism at work, they say.

The 18 researchers, headed by Chris A. Collins of Liverpool John Moores University, thus made a significant discovery confirming the top-down theory of galaxy formation. An article about it in *Science News* is entitled, "Heavyweight Galaxies In The Young Universe - Newfound massive galaxies may force theorists to revisit formation model" [29].

What factors might have caused the rapid growth rate of massive galaxies during their first five billion years, other than through hierarchical galaxy mergers? The factors that we would be addressing would fall into the category of the top-down theory of galaxy formation since the alternative bottom-up theory of galaxy formation is limited to the hierarchical galaxy merger process.

Since dark matter is estimated to represent about 83 percent of the mass of the universe and the mass of the dark matter halo of a spiral galaxy is about ten times greater than the galaxy's ordinary mass, it is logical to search for answers to the rapid-galaxy-growth enigma within the realm of dark matter. We might have tried to work with Cold Dark Matter WIMPS (Weakly Interacting Massive Particles), but since their makeup is unknown and their existence has not been confirmed, even after 25 years of research, analysis could not yield much new information that could be used to solve the enigma of the early rapid-galaxy- growth observations.

Fortunately, there is a promising form of dark matter that we can try to use. It is called relativistic-proton dark matter, which was first described by Bell Labs-trained scientist Jerome Drexler in his December 2003 book, *How Dark Matter Created Dark Energy and the Sun*[34]. In this book he plausibly explains the identity and nature of the dark matter of the universe and how it relates to the accelerating expansion of the universe and to the source of the highest-energy cosmic ray protons observed in the universe. Although his 2003 book was scientifically radical, no scientist has yet written a scientific paper, book, or article opposing Drexler's relativistic-proton dark matter model.

This chapter will now proceed to provide a logical and plausible explanation for the recently discovered high growth rate of massive galaxies during the first five billion years after the big bang. The explanation relies on the dark matter of the universe being comprised primarily of relativistic-proton dark matter; the existence of which is supported by Drexler's books, *Comprehending and Decoding the Cosmos*[35], 2006, and *Discovering Postmodern Cosmology*[36], 2008, and scientific papers in 2005 and 2007.

For almost all of the last twenty years, cosmology professors taught that massive galaxies were formed by a multi-step process. The gravitational collapse of small gas clumps would occur first, followed by their gravitational merger into larger and larger galaxies. This multi-step merger process is called a hierarchical galaxy formation process, which is in the category of the bottom-up theory of galaxy formation, for obvious reasons. Drexler may have been one of the first scientists in recent years to question the hierarchical galaxy formation process and its associated bottom-up theory in his paper, astro-ph/0504512, April 22, 2005, "Identifying Dark Matter through the Constraints Imposed by Fourteen Astronomically Based 'Cosmic Constituents'"[15]. A relevant paragraph from page 9 of this paper reads:

"Mature galaxies in a young Universe: The recent discovery of the existence of mature galaxies only about 2.5 billion years after the Big Bang [30,31,32] (and confirmed by the Carnegie Observatories on March 10, 2005) can be explained using the relativistic proton dark matter theory that involves fast protons that slow down over time due to synchrotron radiation losses, but raises questions about the cold dark matter bottom-up theory of galaxy formation which involves only slow-moving particles. The referenced articles in the July 2004 issue of *Nature* are entitled, 'A high abundance of massive galaxies 3-6 billion years after the Big Bang'[30] and 'Old galaxies in the young Universe'[31]. The Carnegie Observatories had announced in a news release on March 10, 2005 that 'Astronomers have found distant red galaxies "very massive and old" in the universe when it was only 2.5 billion years post Big Bang.'"

The 2009 *Nature* paper involving early galaxy growth is more advanced than the July 2004 early-galaxy-growth *Nature* paper in several respects. The 2009 *Nature* paper sought explanations based upon the top-down theory of galaxy formation, the existence of the filamentary dark matter web passing through galaxies (and galaxy clusters), and the possible role of filamentary dark matter as a pipe or conduit of hydrogen into large and massive galaxies that were experiencing a high rate of growth during the first 5 billion years after the big bang. The 2009 *Nature* paper might have been more advanced if it had included both the earlier referenced paragraph and the following paragraph from Drexler's April 22, 2005 paper:

"Long, large DM filaments creating galaxy clusters: The September 8-9, 2004 news releases from NASA/Harvard entitled, 'Motions in nearby galaxy cluster reveal presence of hidden superstructure'[33] regarding Chandra x-ray images of the Fornax cluster states: 'Astronomers think that most of the matter in the universe is concentrated in long large filaments of dark matter and that galaxy clusters are formed where these filaments intersect.' It should be noted that such a filamentary dark matter structure could be a slightly curved portion of a DM halo around or within some galaxy supercluster. This relatively new top-down theory of galaxy cluster formation is compatible with the relativistic proton dark matter theory as described in the author's book published in December 2003 [34]. (Prior to September 8, 2004, the standard theory of cold dark matter galaxy formation was based upon the bottom-up hierarchical model wherein small galaxies form first and then gravitationally move together over time to form larger galaxies and galaxy clusters.)"

Thus, from a cosmology standpoint, Drexler's April 2005 paper might be considered more advanced than the April 2009 *Nature* paper, "Early assembly of the most massive galaxies"[28]. This would be based upon Drexler's four-year-earlier recognition of (1) the importance of the top-down theory of galaxy formation, (2) the importance of filamentary dark matter to act as a pipe or conduit of hydrogen into large and massive galaxies that were experiencing a high rate of growth during the first 5 billion years after the big bang, (3) the importance of his relativistic-proton dark matter in being able to both provide and feed protons/hydrogen fuel directly into galaxies and galaxy clusters, (4) the importance of his relativistic-proton dark matter in creating muons and in triggering and facilitating the hydrogen fusion in stars, and

(5) the importance of his relativistic-proton dark matter in causing the accelerating expansion of the universe and in providing the source of dark energy.

CHAPTER 12

Drexler's Dark Matter Essentially Predicts Lyman-Alpha Blob, Himiko, Just Discovered by Carnegie Institution

May 18, 2008 — A group of recent science articles about the discovery of the most distant and largest Lyman-alpha blob, dubbed Himiko, has the intriguing titles "Astronomers discover ancestors of modern-day spiral galaxies"[37], "Mysterious Space Blob Discovered at Cosmic Dawn"[38], and "Experts Puzzled by Strange Space Blob"[39]. They all relate to a scientific paper entitled, "Discovery of a Giant Lyα Emitter Near the Reionization Epoch" [40], authored by a group of 27 researchers led by Pasadena's Observatories of the Carnegie Institution of Washington and published May 10, 2009 in The *Astrophysical Journal*. The Carnegie Institution's April 22 news release quotes a team member.

"One of the puzzling things about Himiko is that it is so exceptional," said Carnegie's Alan Dressler, a member of the team. "If this was the discovery of a class of objects that are ancestors of today's galaxies, there should be many more smaller ones already found — a continuous distribution. Because this object is, to this point, one-of-a-kind, it makes it very hard to fit it into the prevailing model of how normal galaxies were assembled. On the other hand, that's what makes it interesting."

The Carnegie Institution news release also conjectures about the nature of Himiko. "It could be ionized gas powered by a super-massive black hole; a primordial galaxy with large gas accretion; a collision of two large young galaxies; super wind from intensive star formation; or a single giant galaxy with a large mass of about 40 billion Suns."

To Bell Labs-trained scientist Jerome Drexler, the Himiko giant Lyman-alpha blob is a visual manifestation of and support for his relativistic-proton filamentary dark matter model linked to his top-down theory of galaxy formation. Of the Carnegie Institution's five possible descriptions listed above, Himiko seems to best fit, "a primordial galaxy with large gas accretion." More specifically, Lyman-alpha blobs are probably partially comprised of Cosmic Web filaments of Drexler's relativistic-proton dark matter, which he discovered in early 2002 and disclosed in his December 2003 book, *How Dark Matter Created Dark Energy and the Sun* [34].

The large volumes of electrons necessary to convert dark matter's relativistic protons into hydrogen, thereby producing the blob's Lyman-alpha emission line photons, could be created by non-elastic collisions of dark matter's relativistic protons with photons from several possible sources. This could produce, in close proximity to the protons, large volumes of intimate pions, which quickly decay into muons, then into intimate electrons to transform the protons into hydrogen, thereby producing the Lyman-alpha emission.

Thus, Drexler's relativistic-proton dark matter model, in conjunction with his top-down theory of galaxy formation, automatically solves the "mystery" of Himiko and essentially predicts it. On the other hand, mainstream cosmologists, who support the bottom-up theory of hierarchical galaxy formation through galaxy mergers, would predict the non-existence of Lyman-alpha blobs and the non-existence of Himiko. Drexler posited his top-down theory of galaxy formation and his relativistic-proton filamentary dark matter model in his scientific paper, astro-ph/0504512, April 22, 2005, "Identifying Dark Matter through the Constraints Imposed by Fourteen Astronomically Based 'Cosmic Constituents' ", which is available at http://www.jeromedrexler.org .

Three years after this paper was published, scientific papers began to be published supporting Drexler's top-down theory of galaxy formation by the following senior staff/institutions: Michael J. Disney, Cardiff University, October 23, 2008, "Galaxies appear simpler than expected" [5] in *Nature*; Avishai Dekel, Hebrew University of Jerusalem, January 22, 2009, "Cold streams in early massive hot haloes as the main mode of galaxy formation"[19] in *Nature*; Cheng-Jiun Ma and Harald Ebeling, University of Hawaii, March 10, 2009, "An X-ray/Optical Study of the Complex Dynamics of the Core of the Massive Intermediate-Redshift Cluster MACSJ0717.5+3745"[47] in *The Astrophysical Journal*; and Chris A. Collins, Liverpool John Moores University, April 2, 2009, "Early assembly of the most massive galaxies"[28] in *Nature*.

A Lyman-alpha blob is a huge concentration of protons, electrons, and hydrogen gas, in the early universe, emitting the Lyman-alpha emission line (http://en.wikipedia.org/wiki/Lyman-alpha_line). The Lyman-alpha emission line in the ultraviolet at a wavelength of 121.6 nanometers before being redshifted, is produced by the combining of electrons with ionized hydrogen atoms (protons). These Lyman-alpha blobs are some of the largest known individual objects in the universe.

Some of these gaseous structures are more than 400,000 light years across. So far they have only been found in the high-redshift (http://en.wikipedia.org/wiki/Redshift) universe because of the ultraviolet nature of the Lyman-alpha emission line. Since Earth's atmosphere is very effective at filtering out UV photons, any Lyman-alpha photons observed with Earth-based telescopes would be significantly redshifted. (The Hubble Space Telescope, upgraded with the

UV Cosmic Origins Spectrograph in May 2009 could possibly be used in the future to study Lyman-alpha blobs.).

Himiko is a newly discovered gas/plasma cloud that predates similar Lyman-alpha blobs (http://en.wikipedia.org/wiki/Lyman-alpha_blob) by about two billion years. It is 12.9 billion light years from Earth. An object 12.9 billion light-years away is seen as it existed 12.9 billion years ago, and the light is just now arriving. It appears to have about 10 times more mass than the next largest object found in the early universe, or roughly the equivalent mass of 40 billion suns. At 55,000 light years in diameter, it has about half the diameter of our Milky Way galaxy. Let us now consider in more detail how Drexler's 2005 relativistic-proton filamentary dark matter model can provide a logical and plausible explanation for the Himiko Lyman-alpha blob.

The following four paragraphs introduce the astronomers' 2004 discovery of slightly curved dark matter filaments that form galaxy clusters where the curved dark matter filaments intersect/collide. Today it is widely accepted that the curved dark matter filaments also weave a Cosmic Web [33,43,77] of dark matter filaments that is integral to all galaxy clusters and galaxies — the Lyman-alpha blobs should be no exception.

Drexler's relativistic-proton dark matter is the only dark matter model that (1) can form into slightly-curved, width-confined, long filaments of dark matter, (2) can generate electrons intimate with the protons via pions and thus produce hydrogen within the dark matter proton filaments, (3) can generate the Lyman-alpha UV line, and (4) can form the filamentary Cosmic Web that feeds the generated hydrogen atoms into the galaxies. This group of processes fits Drexler's top-down theory of galaxy formation very well.

These four features of relativistic-proton dark matter are also compatible with both the observed characteristics of dark matter and observed characteristics of Lyman-alpha blobs.

For almost all of the last twenty years, cosmology professors have taught that massive galaxies were formed by the gravitational collapse of small gas clumps followed by their gravitational merger into larger and larger galaxies. This is called a hierarchical galaxy formation process, which is in the category of the bottom-up theory of galaxy formation.

Drexler may have been one of the first scientists in recent years to question the hierarchical galaxy formation process. He did so by positing his top-down theory of galaxy formation in his scientific paper, astro-ph/0504512, April 22, 2005, "Identifying Dark Matter through the Constraints Imposed by Fourteen Astronomically Based 'Cosmic Constituents'". On pages 8 and 13 we find the following two paragraphs to be very relevant to this discussion.

From page 8 of Drexler's 2005 scientific paper: "Long, large DM filaments creating galaxy clusters: The September 8-9, 2004 news releases from NASA/Harvard entitled, 'Motions in nearby galaxy cluster reveal presence of hidden superstructure' regarding Chandra x-ray images of the Fornax cluster states: 'Astronomers think that most of the matter in the universe is concentrated in long large filaments of dark matter and that galaxy clusters are formed where these filaments intersect.' It should be noted that such a filamentary dark matter structure could be a slightly curved portion of a DM halo around or within some galaxy supercluster. This relatively new top-down theory of galaxy cluster formation is compatible with the relativistic proton dark matter theory as described in the author's book published in December 2003."

From page 13 of Drexler's April 2005 scientific paper: "[Relativistic proton dark matter particles could] be concentrated in the long large curved filaments of dark matter (announced by NASA 9/8/04), which form galaxy clusters where the DM filaments intersect. See SigChar P. W, and X. Some relativistic dark matter protons are concentrated in curved long, large dark matter filaments owing to the high relativistic velocities of the protons and to the magnetic fields created by the astrophysical dynamo effect. The author believes that the DM filaments may be slightly curved portions of supercluster halos of DM protons, the widths of which [DM filaments] are confined electrostatically by the presence, within the filaments, of proton-produced [negatively-charged] muons and electrons (muons decay into electrons, etc.). Further, the crashing of intersecting DM filaments could lead to debris of relativistic protons at various energies and electrons from muon decay and slower moving hydrogen, helium and protons — all the necessary ingredients to form galaxy clusters, galaxies and stars."

The previous two paragraphs from Drexler's April 22, 2005 scientific paper support arguments that relativistic-proton dark matter in a Lyman-alpha blob (1) can form into slightly-curved, width-confined, long filaments of dark matter, (2) can generate intimate electrons thereby facilitating the creation of hydrogen within relativistic-proton filaments, (3) can generate the Lyman-alpha UV line [by the means described in item 2], and (4) can form the filamentary Cosmic Web to feed the generated hydrogen into all the galaxies, which precisely fits the top-down theory of galaxy formation. These four features of relativistic-proton dark matter are compatible with both the observed characteristics of dark matter and the observed characteristics of Lyman-alpha blobs.

The slight curvature of the dark matter filaments comes about because the kinetic energy of the relativistic protons in conjunction with the extragalactic magnetic field determine the radius of curvature of the protons' paths, according to the well-known Larmor radius equation. The width-confinement of the dark matter filaments comes about because the widths are confined electrostatically by the presence, within the proton filaments, of negatively-charged muons and electrons, which reduce the radial outward coulomb forces on the relativistic-proton filamentary stream. Both of these subjects are discussed in the above paragraph taken from page 13 of Drexler's April 22, 2005, scientific paper.

The formation of Lyman-alpha blobs is dependent upon the four cosmic features, mentioned above, as well as the creation of intimate pions through non-elastic collisions of relativistic-protons and photons. These intimate pions quickly decay into negative muons and then into intimate electrons, which combine with the relativistic protons to form hydrogen and generate the Lyman-alpha UV emission line. The protons and hydrogen would be conducted to a Lyman-alpha blob, such as Himiko, through the filamentary dark matter Cosmic Web, as per the top-down theory of galaxy formation. The Lyman-alpha emission line from Lyman-alpha blobs and from Himiko would be significantly red-shifted.

To create the necessary large volumes of pions that would decay into the electrons, which would then combine with dark matter's relativistic protons to form hydrogen and produce the Lyman-alpha emission line, some large supply of intimate photons would be required to enter into non-elastic collisions with the dark matter's relativistic protons, thereby creating pions. What photons are available for this task?

The obvious intimate photons that might possibly fit these pion-source requirements would include (1) the Cosmic Microwave Background (CMB) photons, (2) synchrotron emission photons generated by dark matter's relativistic protons racing through the extragalactic magnetic field, (3) the Lyman-alpha UV emission-line photons created in the early universe by the combining of electrons with protons to form hydrogen atoms, and (4) the muon-based Lyman-alpha X-ray emission-line photons created in the early universe by the combining of negative muons with protons to form short-lived muonic hydrogen atoms.

Note that the mass of a negative muon is about 209 times the mass of an electron, which leads to the Lyman-alpha X-ray emission line for muons as compared to the ultraviolet Lyman-alpha emission line for electrons. More specifically, Aldo Antognini researching muonic hydrogen in Switzerland in 2005, reported, "The subsequent de-excitation to the 1S state emits a 1.9 keV Lyman-alpha x-ray," which corresponds to a 0.65 nanometer X-ray photon.

CHAPTER 13

Drexler's Dark Matter and Dark Energy Boosted by NASA's Choice of Ultraviolet Spectrograph for Hubble

May 27, 2009 — NASA/Hubble's recent choice of the ultraviolet (UV) Cosmic Origins Spectrograph to search and analyze filaments of dark matter in the Cosmic Web [33, 43,77] was hailed today by Bell Labs-trained scientist Jerome Drexler. Drexler had announced his discovery of the only dark matter model that emits ultraviolet photons and explains dark energy in his 2003 book, *How Dark Matter Created Dark Energy and the Sun*[34] and scientifically in a 19-page paper, astro-ph/0504512 [15], April 22, 2005, "Identifying Dark Matter through the Constraints Imposed by Fourteen Astronomically Based 'Cosmic Constituents' ".

To provide overwhelming evidence proving the validity of his dark matter discovery, Drexler utilized it to solve two dozen known cosmic mysteries via his two subsequent books; one published in 2006 entitled *Comprehending and Decoding the Cosmos*[35] and another in 2008 titled *Discovering Postmodern Cosmology* [36]. (See Drexler's *Dark Matter Cosmology* Web site http://www.jeromedrexler.org.)

Drexler's relativistic-proton dark matter emits UV photons directly through synchrotron radiation (http://en.wikipedia.org/wiki/Synchrotron_radiation) since its protons race across extragalactic magnetic field lines. It also can create UV photons indirectly when its protons enter into non-elastic collisions with photons thereby producing close-proximity pions that quickly decay (http://en.wikipedia.org/wiki/Pion) into muons (http://en.wikipedia.org/wiki/Muon) and then into intimate electrons, which can combine with the protons in the dark matter stream to produce hydrogen and the Lyman-alpha (http://en.wikipedia.org/wiki/Lyman-alpha_line) UV photon emission line at 122 nanometers (nm).

Joel Achenbach of the *Washington Post* recently reported on the May 16, 2009 Hubble Space Telescope repair mission, "The astronauts replaced COSTAR with an [ultraviolet] instrument called the Cosmic Origins Spectrograph (COS), which is designed to probe the fundamental structure of the universe and search for filaments of dark matter that bind together the galaxies and all other visible matter" [41]. Dennis Overbye of the *New York Times* wrote, "… [A]stronomers hope to trace a 'cosmic web' of gas and dark matter [with UV] that stretches through the universe connecting galaxies like knots."

David Perlman, SF Chronicle Science Editor wrote an article on May 24, 2009 entitled, "Hubble probing mysteries of deep space"[42]. He reported on an interview of Mario Livio, a well-known astrophysicist, regarding the recent upgrade of the Hubble Space Telescope and on the significance of the installation of the ultraviolet Cosmic Origins Spectrograph (COS). Several of Livio's comments were relevant to examining what is known as the Cosmic Web [33,43], which is formed by the crisscrossing of large, long filaments of dark matter that create galaxy clusters where the dark matter filaments collide. Livio's several informative comments about the Hubble's Cosmic Origins Spectrograph follow:

"It will let us examine the great cosmic webs we cannot see," Livio said.

"Those webs are great regions of intergalactic space that are so thin and tenuous they can never be seen from Earth. Where they are backlit by the brilliance of hugely energetic quasars, however, Hubble will be able to change that."

It will "reveal the structure of the immense filaments inside the webs we cannot see," Livio said.

The Cosmic Origins Spectrograph was installed on the Hubble Space Telescope (HST) [44] on May 16, 2009. It is designed to perform high sensitivity, moderate-resolution and low-resolution spectroscopy of astronomical objects in the 115-320 nanometer ultraviolet (UV) wavelength range. COS will significantly enhance the spectroscopic capabilities of HST at UV wavelengths, and will provide observers with unparalleled opportunities for observing faint sources of UV light. One goal of COS is to identify dark matter, which has been one of science's greatest mysteries since it was detected in the 1930's by Fritz Zwicky[45] then of Caltech.

Ever since the galaxies-orbiting relativistic-proton dark matter model was developed by Drexler over seven years ago (and the filamentary dark matter model over four years ago), he has been aware that his dark matter model should emit synchrotron radiation of ultraviolet photons and possibly extreme ultraviolet (EUV) photons, and soft X-ray photons. He also has recognized the possibility of the Lyman-alpha UV photon emission line at 122 nm and the muon-based Lyman-alpha soft X-ray emission line at 0.65 nm. After Drexler published his 2003 book, astronomers doubting his relativistic proton dark matter model had argued that if his model were correct, ultraviolet photons emitted from dark matter orbiting groups of galaxies would have been detected by astronomers years ago.

Drexler had responded that the failure of astronomers to detect the UV synchrotron emission from the relativistic-proton dark matter probably was caused by the low power level of the ultraviolet photon emission and the absorption by Earth's atmosphere which is very effective in filtering out UV photons. Thus, to detect UV from relativistic-proton dark matter probably would require a satellite UV observatory. Therefore, NASA's installation of the UV

Cosmic Origins Spectrograph on May 16 was a dream come true.

The probability of detecting ultraviolet photons from relativistic-proton dark matter rose on June 20, 2006, when Russia announced that it will launch the Spektr-UF ultraviolet astronomical observatory, which would have a 1.7 meter diameter main mirror, into a highly elliptical orbit in 2010 [46]. In the news release, the Russian project manager and Professor of Physics and Mathematics Boris Shustov, who heads the Institute of Astronomy of the Russian Academy of Sciences, is quoted as saying, "One should particularly emphasize the observatory's role in detecting the so-called dark matter of the Universe and unlocking its secrets because such dark matter can only be seen by large ultraviolet telescopes."

Thus, Professor Shustov's statement, that "dark matter can only be seen by large ultraviolet telescopes" and NASA/Hubble's installation of the UV-based COS to detect dark matter in the Cosmic Web, both boost Drexler's relativistic-proton dark matter model and its closely related dark energy model.

Let us go back to Hubble and the UV COS. In order to give the public some insight into Hubble's UV Cosmic Origins Spectrograph, NASA organized a conversation on the subject between two of its scientific leaders and posted it on a NASA Web site on September 2, 2008. The conversation flowed as follows:

Dave Lekrone, HST Chief Scientist, NASA: "Spectroscopes or spectrographs are absolutely essential in that toolbox of astronomical tools that are so important for research. They produce ugly pictures. But they are the nuts and bolts of physical science. They put the physics in astrophysics.

"If you look out across the universe today and you start seeing this inhomogeneous web-like structure with filaments, places where filaments come together, looks just like a big three-dimensional spider web, tracing all those filaments is the light of ordinary stars and galaxies."

Mike Shull, Astronomer, University of Colorado: "It's ordinary matter that the planets and stars and humans are made of, hydrogen, helium, chemical elements. But it's been a mystery because galaxies and stars account for maybe just 10 percent of this matter, ordinary matter in the universe. We've been wondering for decades, 'Where's the rest of this matter?'.

"With dark matter almost being certainly have been produced as a byproduct of the Big Bang, all that gravity from the dark matter tended to be the most important force that pulled material together."

Dave Lekrone, HST Chief Scientist, NASA: "COS wants to trace that part of the story, the history of our universe."

Mike Shull, Astronomer, University of Colorado: "There's two things you want to do as an astronomer. You want to build a bigger telescope to gather more light and you also want to spread it out to its colors, its components. It's called taking a spectrum. COS does both of those things 10 or 20 times better than has ever been done before."

Dave Lekrone, HST Chief Scientist, NASA: "If you want to know what something is made of, how hot it is, how dense it is, how fast it's moving in space. A spectrograph will give you all that information. With COS, we can acquire information like that, farther out across the universe than we've ever been able to do before. Deep down inside, we just want to know where we came from and how we got here."

CHAPTER 14

University of Hawaii Astronomers Boost Drexler's Dark Matter and Galaxy Formation Models Published in 2005

June 11, 2009 — Cheng-Jiun Ma, Harald Ebeling, and Elizabeth Barrett at the Institute of Astronomy of the University of Hawaii recently published a scientific paper in the *Astrophysical Journal Letters* (ApJ 693, L56-L60), which reports, "we also find tantalizing, if circumstantial, evidence for direct, large-scale heating of the ICM [intracluster medium] by contiguous infall of low-density [hydrogen] gas from the [dark matter] filament."

Their paper entitled, "An X-ray/Optical Study of the Complex Dynamics of the Core of the Massive Intermediate-Redshift Cluster MACSJ0717.5+3745" [47] was published on March 10, 2009. *New Scientist* published a May 4, 2009 article about this scientific paper, entitled "Dark matter 'highway' funnels gas into galactic pileup" [48], which provides very useful information that is quoted as follows: "and researchers say as much as 40 percent of all dark matter in the universe may lie in the [dark matter] filaments."

"Now, he [Harald Ebeling] and a team led by Cheng-Jiun Ma say they see hints of [hydrogen] gas in a [dark matter] filament that appears to be funneling galaxies into a galaxy cluster already crammed with them."

In response to these intriguing observations, "The team has submitted a proposal to observe the [galaxy] cluster again — for a longer period — with the Chandra X-ray Observatory."

"They also hope to see [hydrogen] gas within the [dark matter] filament that lies farther away from where the filament connects with the cluster. 'Right now, we're only just barely seeing (filament gas) where it hits the cluster,' he [Ebeling] says."

"'But as you move away from the cluster, the [hydrogen] gas is still unperturbed,' Ebeling says 'This is the pristine, original state of [hydrogen] gas in [dark matter] filaments …'"

The researchers' scientific paper and related astronomical research at the University of Hawaii add support to both the dark-matter-filament model and top-down theory of galaxy formation of Bell Labs-trained scientist Jerome Drexler, posited in his April 22, 2005 scientific paper [15].

Drexler's 2005 top-down galaxy formation model did not gain traction until the 2009 academic year when it received strong support from researchers at Cardiff University of Wales [5], Hebrew University of Jerusalem [19], University of Hawaii [47] [48] and the Liverpool John Moores University [28].

For almost all of the last twenty years, cosmology/ astrophysics professors have taught that massive galaxies were formed by the gravitational collapse of small gas clumps followed by their gravitational merger into larger and larger galaxies. This is called a hierarchical galaxy formation process, which is in the category of the bottom-up theory of galaxy formation.

Drexler may have been one of the first scientists in recent years to question the hierarchical galaxy formation process. He did so by positing his top-down theory of galaxy formation theory (and dark matter filament theory) in his scientific paper, astro-ph/0504512v1[15], April 22, 2005, "Identifying Dark Matter through the Constraints Imposed by Fourteen Astronomically Based 'Cosmic Constituents'". On pages 8 and 13 we find the following two paragraphs that explain both Drexler's dark-matter-filament model and his top-down theory of galaxy formation, which essentially predict and explain the recent astronomical discovery of Ma, Ebeling, and Barrett, reported in their paper.

A paragraph on page 8 of the 2005 paper reads:

> "Long, large DM filaments creating galaxy clusters. The September 8-9, 2004 news releases from NASA/Harvard entitled, 'Motions in nearby galaxy cluster reveal presence of hidden superstructure,' regarding Chandra x-ray images of the Fornax cluster states: 'Astronomers think that most of the matter in the universe is concentrated in long large filaments of dark matter and that galaxy clusters are formed where these filaments intersect.' It should be noted that such a filamentary dark matter

structure could be a slightly curved portion of a DM halo around or within some galaxy supercluster. This relatively new top-down theory of galaxy cluster formation is compatible with the relativistic proton dark matter theory as described in the author's book published in December 2003."

A paragraph on page 13 of Drexler's 2005 scientific paper reads:

"[Relativistic proton dark matter particles could] be concentrated in the long large curved filaments of dark matter (announced by NASA 9/8/04), which form galaxy clusters where the dark matter filaments intersect. Some relativistic dark matter protons are concentrated in curved long, large dark matter filaments owing to the high relativistic velocities of the protons and to the magnetic fields created by the astrophysical dynamo effect. The author believes that the dark matter filaments may be slightly curved portions of supercluster halos of dark matter protons, the widths of which [dark matter filaments] are confined electrostatically by the presence, within the filaments, of proton-produced [negatively-charged] muons and electrons (muons decay into electrons, etc.). Further, the crashing of intersecting dark matter filaments could lead to debris of relativistic protons at various energies and electrons from muon decay and slower moving hydrogen, helium and protons — all the necessary ingredients to form galaxy clusters, galaxies and stars."

The previous two paragraphs from Drexler's April 22, 2005 scientific paper explain how the top-down theory of galaxy formation is intimately linked to the relativistic-proton filamentary dark matter model, which (1) can form into curved, width-confined, long filaments of dark matter, (2) can generate electrons, via the creation of pions (see below) (http://en.wikipedia.org/wiki/Pion) that decay into muons (http://en.wikipedia.org/wiki/Muon) that decay into electrons, thereby facilitating the production of hydrogen gas within relativistic-proton dark matter filaments, (3) can generate the Lyman-alpha UV photon emission line [by the means described in item 2], and (4) can form the filamentary dark matter Cosmic Web [33,43,77] to feed the generated hydrogen gas into the galaxies.

These four features of relativistic-proton filamentary dark matter are compatible with and facilitate the top-down galaxy formation process posited in 2005 and astronomically discovered in 2009 by the three University of Hawaii astronomers.

The curvature of the relativistic-proton dark matter filaments comes about because the kinetic energy of the relativistic protons in conjunction with the extragalactic magnetic field determine the radius of curvature of the protons' looping paths, according to the well-known Larmor radius equation. The Cosmic Web structure of the looping dark matter filaments can be explained only by a relativistic-proton dark matter stream passing through a slightly non-uniform, extragalactic magnetic field. The width-confinement of the dark matter filaments comes about because the widths are confined electrostatically by the presence, within the proton filaments, of negatively-charged muons and electrons, which reduce the radial outward coulomb forces on the relativistic-proton filamentary stream. Both of these subjects are discussed in the above paragraph taken from page 13 of Drexler's April 22, 2005 paper.

Drexler's top-down galaxy formation model is dependent upon the four cosmic features listed above, which includes the creation of intimate pions through non-elastic collisions of relativistic-protons with photons. These intimate pions quickly decay into negative muons then into intimate electrons, which combine with the relativistic protons to form hydrogen. This hydrogen gas and the relativistic protons would be conducted to the galaxies via the filamentary dark matter Cosmic Web.

To create the necessary large volumes of pions that would decay into muons and then into electrons, which would then combine with dark matter's relativistic protons to form hydrogen, some large supply of close-proximity photons would be required to enter into non-elastic collisions with the dark matter's relativistic protons. What intimate photons are available for this task?

The obvious intimate photons that might possibly fit these pion-source requirements would include (1) the Cosmic Microwave Background (CMB) photons, (2) synchrotron emission photons generated by dark matter's relativistic protons racing through the extragalactic magnetic field, (3) the Lyman-alpha UV emission-line photons, redshifted to various degrees, created by the combining of electrons with protons to form hydrogen atoms, (4) the muon-based Lyman-alpha X-ray emission-line photons, redshifted to various degrees, created by the combining of negative muons with protons to form short-lived muonic hydrogen atoms, and (5) photon emission, redshifted to various degrees, through the formation of helium and the short-lived muonic helium, which is possible because for every ten to twelve protons of relativistic-proton dark matter there is approximately one relativistic helium nucleus.

CHAPTER 15

Scientific American: Dark Matter Is Not Made of Protons; Jerome Drexler: Dark Matter is made of Relativistic Protons

June 17, 2009 — An article, entitled "The Search for Dark Matter" in the recent *Majestic Universe* [49] special issue of *Scientific American*, makes the following statements about the dark matter of the universe:

> "[1] What kind of particle could dark matter be made of? Astronomical observation and theory provide some general clues. [2] It cannot be protons or neutrons or anything that was once made of protons or neutrons, such as massive stars that became black holes. [3] According to calculations of particle synthesis during the big bang, such particles were simply too few in number to make up the dark matter. [4] Those calculations have been corroborated by measurements of primordial hydrogen, helium, and lithium in the universe."

Almost identical versions of these four statements were also published in an article with the same title in the March 2003 issue of *Scientific American*. Bell Labs-trained scientist Jerome Drexler indicated in his December 2003 book, *How Dark Matter Created Dark Energy and the Sun,* that these four statements are true only if the protons are moving slowly, but are not true if the protons are moving near the speed of light. That is because Einstein's Special Theory of Relativity says that the mass of such very fast protons could be as much as 100, 1000, or even a million times greater than the mass of slow-moving protons.

Now let us consider sentence [3] above which says "such [proton or neutron] particles were simply too few in number to make up the [actual mass of] dark matter." However, from Einstein's Special Theory of Relativity we see that even if the relativistic protons are "few in number", their very high relativistic mass could still "make up the [actual mass of] dark matter." Drexler had discovered this possibility in early 2002.

His confidence in the relativistic-proton dark matter idea was soon enhanced when he realized that the same concept could also explain the mysterious relativistic cosmic-ray protons bombarding Earth every day that probably represent energy-depleted relativistic-proton dark matter "brought down to Earth".

Drexler became almost completely convinced of the relativistic-proton dark matter idea later in 2002 when this same concept also provided a plausible explanation for the accelerating expansion of the universe, which had been discovered astronomically in 1998.

Drexler's explanation for it requires only basic astrophysics. It has been known through astronomy that spiral disk galaxies are enclosed by a spheroidal halo of dark matter with a mass about ten times greater than the ordinary mass of the spiral galaxy and also that a weak magnetic field pervades the universe.

It is also well known that when relativistic protons cross magnetic field lines in space they lose relativistic mass through the emission of large quantities of photons. This is called synchrotron radiation or emission. Furthermore, it has been known through Astronomer Edwin Hubble for many years that all galaxy clusters are moving away from one another. Using all of this information, let us consider three galaxy clusters that are moving away from each other at some separation velocities, at a certain time.

Let us look again a month later. The masses of all three galaxy clusters would be lower because of their continuous synchrotron emission loss of photons and thus the tidal gravitational attraction between pairs of them would be continuously lowered. For this same reason, all the separation velocities between galaxy clusters in the universe would continue to increase, which is an accelerating expansion of the universe!

By 2004, Drexler's dark-matter-synchrotron-emission model appeared to be the only posited plausible published explanation for the 1998 discovery of the accelerating expansion of the universe. There is still no other published plausible explanation.

In late 2002, with Drexler's relativistic-proton dark matter model solving the three puzzles of the insufficient proton mass in the universe, the relativistic cosmic-ray protons demonstrating relativistic-proton dark matter being "brought down to Earth", and the apparent solving of the mystery of the accelerating expansion of the universe (attributed earlier to a dark energy), he thought about publishing his findings.

The dark matter article in the March 2003 issue of *Scientific American* convinced Drexler that he was on the right track. He then proceeded to author a paperback book, entitled *How Dark Matter Created Dark Energy and the Sun*[34], that was published in December 2003.

To demonstrate the validity and significance of his relativistic-proton dark matter model, Drexler used this model to solve two dozen cosmic mysteries and published the results in two more paperback books, *Comprehending and Decoding the Cosmos,* in 2006 and *Discovering Postmodern Cosmology,* in 2008, and scientific papers in 2005 and 2007. See also Drexler's Web site at http://www.jeromedrexler.org .

The full titles of Drexler's three paperback books and two scientific papers, providing overwhelming evidence that relativistic-proton dark matter represents the principal constituent of the dark matter of the universe, are as follows:

(1)   Book, March 1, 2008, *Discovering Postmodern Cosmology: Discoveries in Dark Matter, Cosmic Web, Big Bang, Inflation, Cosmic Rays, Dark Energy, Accelerating Cosmos.*

(2)   Scientific paper, physics/0702132, Feb. 15 2007, "A Relativistic-Proton Dark Matter Would Be Evidence the Big Bang Probably Satisfied the Second Law of Thermodynamics".

(3)   Book, May 22, 2006, *Comprehending and Decoding the Cosmos: Discovering Solutions to Over a Dozen Cosmic Mysteries by Utilizing Dark Matter Relationism, Cosmology, and Astrophysics.*

(4)   Scientific paper, astro-ph/0504512, April 22, 2005, "Identifying Dark Matter through the Constraints Imposed by Fourteen Astronomically Based 'Cosmic Constituents'".

(5)   Book, Dec. 15, 2003, *How Dark Matter Created Dark Energy and the Sun: An Astrophysics Detective Story.*

CHAPTER 16

Drexler Leads Dark Matter Identity Race; Queen's University of Canada Gets $10.5 Million

July 1, 2009 — On June 19, 2009 the Canadian government's Canadian Foundation for Innovation (CFI) announced that "Mark Boulay (Physics) and co-investigator Mark Chen (Physics) [of Queen's University] will receive $10,561,628 toward their projects searching for Dark Matter particles [in deep underground mines] and extending the experiments at the Sudbury Neutrino Laboratory" [50, 51].

Dark matter research also can be carried out in space. Seven years of such research in dark matter cosmology by Bell Labs-trained scientist Jerome Drexler has achieved significant positive research results in the dark-matter-identification race against research groups working in deep underground mines in Canada, France, Italy, Spain, United Kingdom, and the United States.

For final dark-matter-identity confirmation, Drexler will be relying on the UV upgraded Hubble Space Telescope and on the planned Russian UV space telescope to detect UV photon emission from his posited relativistic-baryon (protons and helium nuclei) dark matter particles, which have immense relativistic mass that could very well represent about 83 percent of the mass of the universe. Such massive charged particles bombard Earth every day and are well-known as ultra-high-energy cosmic rays.

The competing deep-mine research groups are searching for putative non-baryonic dark matter particles that are devoid of both protons and neutrons, yet are required to represent about 83 percent of the mass of the universe. Such naturally occurring, non-baryonic matter has never been detected by mankind. Thus the deep-mine dark matter researchers face at least two high hurdles; they must not only discover the existence of non-baryonic matter on Earth, which could win them a Nobel Prize, but they must also prove it represents about 83 percent of the mass of the universe for it to be considered the dark matter of the universe.

Since 1984[1], cosmologists have argued that the dark matter of the universe cannot be made of protons or neutrons or anything that was once made of protons or neutrons, such as helium nuclei. They say that calculations of particle synthesis during the big bang indicate that such proton and neutron based particles were simply "too few in number" to make up the enormous estimated mass of dark matter.

However, these researchers did not consider the alternative possibility of particles of relativistic-proton (and helium nuclei) dark matter that, owing to their enormous relativistic particle mass, would not be "too few in number" to make up the enormous mass of dark matter of the universe. Discovering this, Drexler adopted the relativistic-proton dark matter model in 2002 and publicly announced it in late 2003.

Nevertheless, based upon the particle synthesis calculations, mainstream researchers have been searching for putative non-baryonic dark matter in deep underground mines for a decade. To date, they have not found any non-baryonic matter in any form. They continue to search for non-baryonic Cold Dark Matter WIMPs (weakly interacting massive particles), also referred to as neutralinos.

Even if they do find neutralinos, non-baryonic dark matter cosmology seems unable to explain any the following 12 cosmic constituents or cosmic phenomena known to exist in the universe. However, Drexler's relativistic dark matter cosmology provides plausible explanations for each and every one of them, as presented in his three books, two scientific papers, and/or two dozen of scientific articles. See the Web site: http://www.jeromedrexler.org/ .

Cosmic phenomena or cosmic constituents known to exist

1. The accelerating expansion of the universe
2. Dark Energy
3. A dark matter that can exist in the form of spheroid halos around spiral disc galaxies and also in the form of long large slightly curved filaments that form the Cosmic Web.

4. The source of ultra-high-energy cosmic-ray protons

5. How Cosmic Inflation started and then stopped during the post-big bang period

6. The source of the magnetic field that pervades the universe

7. Why most large galaxies formed without galaxy mergers

8. The cause for the early rapid growth of massive galaxies

9. The cause for the stunted mass growth of galaxy clusters

10. How the first stars formed without the availability of hydrogen molecules or dust

11. The basis for the formation of the Lyman Alpha blobs

12. Limitation of the maximum diameter of galaxy superclusters to 430 million light years.

Remarkably, Drexler's relativistic-proton dark matter cosmology explains these 12 cosmic phenomena or cosmic constituents based only upon the laws of physics and known astronomical observations, without employing any assumptions. This implies some forms of relationships exist between the 12 items and relativistic-proton dark matter. Thus, the 12 explanations required for these 12 items represent 12 more high hurdles for non-baryonic dark matter to overcome beyond the two hurdles mentioned previously.

Drexler's relativistic-proton dark matter emits UV photons directly through synchrotron radiation (http://en.wikipedia.org/wiki/Synchrotron_radiation) since its protons race across extragalactic magnetic field lines. It also can create UV photons indirectly when its protons enter into non-elastic collisions with photons thereby producing close-proximity pions that quickly decay (http://en.wikipedia.org/wiki/Pion) into muons (http://en.wikipedia.org/wiki/Muon) and then into intimate electrons, which can combine with the relativistic protons in the dark matter stream to produce hydrogen and the well-known Lyman-alpha (http://en.wikipedia.org/wiki/Lyman-alpha_line) UV photon emission line at 122 nanometers.

Based upon this UV photon emission from Drexler's dark matter, he anticipates the confirmation of relativistic-proton dark matter to occur during 2010 or 2011 via the new UV Cosmic Origins Spectrograph (COS) recently installed on the upgraded Hubble Space Telescope (see Chapter 13) or alternatively via a planned Russian UV space telescope announced for launch in 2010 by the official Russian news source RIA Novosti [46].

CHAPTER 17

Dark Matter's Identity Disclosed to 13 Top Leaders of Paris "Invisible Universe" Symposium

July 6, 2009 — The following leaders of the Paris "Invisible Universe" Symposium [52] have been informed of the precise identity of the dark matter of the universe: George Smoot (Physics Nobel laureate 2006) and David Gross (Nobel 2004), together with Alain Connes (Fields Medal 1982), Abhay Ashtekar, Edmund Berschinger, Francoise Combes, Edward Kolb, Mordehai Milgrom, James Peebles, Jean-Loup Puget, Adam Riess, Leonard Susskind and Edward Wright.

This disclosure of the identity of dark matter follows seven years of dark matter research that was separate and distinct from the mainstream non-baryonic Cold Dark Matter research. The disclosure was triggered last week for two reasons: by the growing support for the top-down galaxy formation model closely linked to the identified relativistic-proton dark matter model, and by the implied admission of the failure of the 25-year-old non-baryonic Cold Dark Matter (CDM) model [1] by the Paris "Invisible Universe" Symposium.

Mainstream cosmology is in a crisis. Dark matter has been researched by mainstream scientists for a quarter of a century and dark energy has been researched for a decade with essentially no progress to report for either. In fact, mainstream CDM cosmology took a step backward recently, by beginning to move away from the bottom-up model of galaxy formation involving hierarchical mergers of smaller galaxies into progressively larger ones. The bottom-up merger model had been one of the pillars of the non-baryonic Cold Dark Matter model.

The bottom-up galaxy merger model is being replaced by the top-down model of galaxy formation where hydrogen gas is fed to growing galaxies through the large long slightly curved filaments of dark matter that comprise the Cosmic Web[33,43,77]. This top-down model was first posited by Bell Labs-trained American scientist Jerome Drexler in a 19-page paper, astro-ph/0504512[15], April 22, 2005, "Identifying Dark Matter through the Constraints Imposed by Fourteen Astronomically Based 'Cosmic Constituents'".

Drexler's 2005 top-down galaxy formation model did not gain traction until the 2009 academic year when it received strong support from researchers at Cardiff University of Wales[5], Hebrew University of Jerusalem [19], University of Hawaii[47] [48] and the Liverpool John Moores University[28].

Apparently, with the failure of the bottom-up galaxy-merger model, mainstream cosmologists are beginning to accept the fact that they are in a crisis and must depart from the path through which they have been trudging for the past 25 years.

That recognition would be good news according to Thomas S. Kuhn author of the classic *The Structure of Scientific Revolutions* [53]. He argues that scientific advancement is not evolutionary, but rather is a series of peaceful interludes punctuated by intellectually violent revolutions. He points out that when a crisis situation in science is finally accepted, the mainstream researchers may be prepared to accept the emergence of new scientific models, representing a new scientific paradigm.

Are the cosmologists ready to accept that mainstream cosmology is in a crisis? The following three short paragraphs, which were part of the introduction to the Paris symposium, "Invisible Universe — Toward a New Cosmological Paradigm"[52], (June 29-July 10), provide some clues:

"Cosmology has arrived at a crossroads. According to the best data available, from large ground-based telescopes and space observatories, almost 95 percent of the universe irretrievably escapes observational detection."

"This missing part of the cosmos is constituted for 25 percent by a mysterious form of dark matter and 70 percent, by a dark energy whose nature is even more exotic and unknown! But what are exactly these new physical entities?"

"In an attempt to answer this complex and profound question, more than 400 experts will gather in Paris to evaluate the situation, and draw future perspectives. The basic principles of physics appear sometimes to be put into question. Modern cosmology is perhaps at the beginnings of a major renewal, similar to those once made by Galileo and Einstein."

How do we proceed to turn around the failing cosmology research? Science writer Ehsan Masood wrote an article, "Are we witnessing the end of science?"[54], which contains two paragraphs suggesting how we might proceed:

> "Revolutions in scientific thinking are always difficult — but perhaps one reason why we may see fewer of them in the future is because of the highly professional way in which modern science is organized. It takes a lot of courage to challenge conventionally accepted views, and it needs a certain amount of stamina to constantly battle those who want to protect the status quo. Mavericks do not do well in large organizations, which is what some scientific fields have become."

> "Progress in science needs researchers who are not afraid — or who are encouraged and rewarded — to ask awkward and difficult questions of theory and of new data. It is easier to question mainstream views if you are independently wealthy, as many scientists in previous ages tended to be. But I wonder how many of us would do so if we were employed by the state and our career progression depended on the validation of our peers?"

Fortunately for scientific progress in cosmology, Bell Labs-trained scientist Jerome Drexler took it upon himself to launch his own self-financed scientific revolution in cosmology in 2002 via what he now calls *dark matter cosmology* or *postmodern cosmology*. He has discovered that the universe is much more orderly, logical and systematic than mainstream scientists have imagined.

For example, starting with relativistic-proton dark matter, the laws of physics, and well-known astronomical data, a capable scientist can predict the existence of ultra-high-energy cosmic rays, the accelerating expansion of the universe (dark energy), the Cosmic Web, the top-down theory of galaxy formation, and cosmic inflation. Drexler coined the expression *dark matter relationism*[35] to describe these multiple cosmic relationships of dark matter with the various constituents and phenomena of the universe.

Mainstream cosmologists have been treating dark matter, ultra-high-energy cosmic rays, the accelerating expansion of the universe, the Cosmic Web, the top-down theory of galaxy formation, and cosmic inflation as if they are all independent of one another. Drexler began to consider possible relationships between dark matter and some of the cosmic constituents and cosmic phenomena when he realized that dark matter represents an amazing 83 percent of the total mass of the universe. This led him to the above-mentioned concept of *dark matter relationism.*

CHAPTER 18

Dark Matter Tops MOND and Creates an Accelerating Cosmos, Cosmic Web, Inflation, Cosmic Rays, Top-Down Galaxy Formation

July 14, 2009 — *New Scientist* magazine published an article[55], based upon a scientific paper[56], about the MOND (modified Newtonian dynamics) gravitational theory entitled, "Phantom menace to dark matter theory". This MOND gravitational theory might have been a *menace* to the simple dark matter gravitational theories of the last century. However, in December 2003 dark matter became the principal feature and key to a new cosmology paradigm, not just a simple gravitational phenomenon, with the publication of a book, *How Dark Matter Created Dark Energy And The Sun* [34], by Bell Labs-trained scientist Jerome Drexler.

This *dark matter cosmology* model was further developed and enhanced in an April 2005 scientific paper by Drexler, which further separates his dark matter model from the simple gravitational model. This paper is astro-ph/0504512[15], April 22, 2005, "Identifying Dark Matter through the Constraints Imposed by Fourteen Astronomically Based 'Cosmic Constituents'". It presents an image of Drexler's relativistic-proton dark matter (83 percent of the mass of the universe) racing through the universe that is as active, dynamic and essential to the universe as the blood-water mass (70 percent of the mass of the human body) flowing through our bodies is essential to life.

Dark matter racing through the universe at relativistic speeds creates or causes the accelerating expansion of the universe (dark energy), the Cosmic Web structure of dark matter filaments, hyper inflation growth of the universe during the big bang period, ultra-high-energy cosmic rays, and the top-down theory of galaxy formation (recently confirmed at Cardiff University of Wales[5], Hebrew University of Jerusalem [19], the University of Hawaii [47] and Liverpool John Moores University[28]). We now know that dark matter is not a passive gravitational phenomenon as supporters of modified Newtonian dynamics (MOND) would have us believe. Neither is it comprised of the weakly interacting non-baryonic passive Cold Dark Matter neutralinos that its supporters would have us believe.

Significantly, both the MOND gravitational theory and the putative non-baryonic Cold Dark Matter theory appear to be too underdeveloped to explain any the following 12 cosmic constituents or cosmic phenomena known to exist in the universe. However, Drexler's relativistic-proton dark matter cosmology provides plausible explanations for each and every one of them, as presented in his three books, two scientific papers, and/or two dozen 2008/2009 scientific articles. See the Web site: http://www.jeromedrexler.org/ .

Cosmic phenomena or cosmic constituents known to exist:

1. The accelerating expansion of the universe

2. Dark Energy

3. A dark matter that can exist in the form of spheroidal halos around spiral disc galaxies and also in the form of long large slightly curved filaments that form the Cosmic Web.

4. The source of ultra-high-energy cosmic-ray protons

5. How Cosmic Inflation started and then stopped during the post-big bang period

6. The source of the universe's magnetic field

7. Why most large galaxies formed without galaxy mergers

8. The cause for the early rapid growth of massive galaxies

9. The cause for the stunted mass growth of galaxy clusters

10. How the first stars formed without availability of molecular hydrogen or dust

11. The basis for the formation of the Lyman Alpha blobs

12. Limitation of the maximum diameter of galaxy superclusters to 430 million light years.

Remarkably, Drexler's relativistic-proton dark matter cosmology explains these 12 cosmic phenomena or cosmic constituents by utilizing only the laws of physics and known astronomical observations, without employing any assumptions. This implies that some form of relationship exists between each of the dozen cosmic constituents or phenomena and relativistic-proton dark matter. Thus, the dozen cosmologic explanations that are required for these 12 items represent 12 more high hurdles for the MOND gravitational theory to overcome and for the putative non-baryonic Cold Dark Matter model to overcome.

CHAPTER 19

Dark Energy Computer Simulations by NASA / U.S. Department of Energy Could Be Based on Drexler's Dark Matter / Dark Energy Link

July 28, 2009 — Recently, Space.com published a well-written article entitled, "NASA Poised to Join Europe's Mars Rover Mission" [57]. However, the first paragraph ends with, "but [they] have agreed to go their separate ways, for now, in exploring the mysteries of dark energy, according to U.S. [NASA / U.S. Department of Energy (U.S. DOE)] and European officials."

Bell Labs-trained discoverer of relativistic-proton dark matter, Jerome Drexler, believes that this delay of cooperation on "exploring the mysteries of dark energy" may be a cost-savings opportunity for NASA and the U.S. DOE. This opportunity is based upon utilizing Drexler's research discoveries in dark matter and dark energy, published in his scientific papers in 2005 and 2007 and in his books in 2003, 2006, and 2008, that could be used to achieve significant dark energy research results with a small expenditure on a computer simulation. (For information about Drexler's seven-year research program see Web site at http://www.jeromedrexler.org .)

Drexler recommends that NASA / U.S. DOE launch a dark energy computer simulation effort related to his linked models of the dark energy and dark matter of the universe. The remainder of this newswire explains how Drexler's seven-year research effort in dark matter and dark energy evolved into this proposed dark energy computer simulation project.

As previously mentioned in this book, Drexler became almost completely convinced of the relativistic-proton dark matter model late in 2002 when this same concept also provided a plausible explanation for the accelerating expansion of the universe, which had been discovered astronomically in 1998.

Drexler's simpler explanation for it proceeds as follows. It has been known through astronomy that spiral disk galaxies are enclosed by a spheroidal halo of dark matter with a mass about ten times greater than the ordinary mass of the spiral galaxy. From astronomy, we also know that a weak magnetic field pervades the universe.

It is also well known that when relativistic protons cross magnetic field lines in space they lose relativistic mass and energy through the emission of a large quantities of photons.

This is called synchrotron emission (or radiation). Furthermore, it has been known through astronomer Edwin Hubble for many years that all galaxy clusters are moving away from one another. Using all of this information, let us consider three galaxy clusters that are moving away from each other at some separation velocities.

Let us check these galaxy clusters a month later. The mass of each of the three galaxy clusters would be lower because of their continuous synchrotron emission loss of photons and thus the tidal gravitational attractive force between pairs of them also would be lower.

Therefore, their separation velocities would be higher and rising. For this same reason, all the separation velocities between galaxy clusters in the universe should be increasing, which describes, in fact, *an accelerating expansion of the universe.*

This briefly describes Drexler's dark energy model / theory, which was published in 2003, 2005, 2006, and 2008. It appears to be the only published plausible explanation for the 1998 discovery of the accelerating expansion of the universe. NASA and the U.S. DOE have not publicly adopted or tested Drexler's accelerating-expansion (dark energy) model. It would seem prudent, plausible, and logical for them to conduct computer simulations of Drexler's dark energy model before launching an expensive new satellite telescope.

The computer models to be used might be based upon well-documented groups of three galaxy clusters for which the mass of each galaxy cluster and the three inter-cluster separation velocities can be determined or estimated.

To help optimize the proposed NASA / DOE dark energy computer simulations, the following more-detailed discussion of Drexler's accelerating-expansion theory/model is provided:

In an expanding universe galaxy clusters are moving away from one another. When the galaxy-cluster separation velocities speed up, as in our universe, the universe is in an accelerating expansion mode. Since 1998, this has been attributed to a mysterious *dark energy*, called "the most profound mystery in all of science" by University of Chicago cosmologist Michael Turner.

In 1929, Edwin P. Hubble announced from the Mount Wilson Observatory near Los Angeles that with the exception of the galaxies closest to the Milky Way, galaxies are rushing away from each other in all directions and, therefore, the universe is expanding. His astronomical evidence led to the big bang theory of the creation of the universe.

In 1998, a completed ten-year study led by Saul Perlmutter, of the Lawrence Berkeley National Laboratory and Brian Schmidt of the Australian National University, of astronomical events involving exploding stars (supernovae) led to the discovery that the expansion of the universe was accelerating. These measurements were made by ground-based telescopes. More recently, Type 1a supernovae studies with the Hubble Space Telescope confirmed and refined the 1998 accelerating-expansion conclusions.

The cause was attributed to *dark energy*, a hypothetical form of energy that may permeate all space and may have negative pressure resulting in a repulsive gravitational force. In 2001, Michael Turner essentially removed the word *all* from this definition when he wrote, "Dark energy by its very nature is diffuse and a low-energy phenomenon. It probably cannot be produced at accelerators; it isn't found in galaxies or even clusters of galaxies." Note that, "it isn't found in galaxies or even clusters of galaxies" actually means there is no evidence of galaxies being pushed apart or stars being pushed apart.

Thus, astronomical data provide three key clues as to the putative mysterious repulsive dark energy force: (1) it pushes galaxy clusters apart; (2) it doesn't push galaxies apart; and (3) it does not push stars apart. Drexler's accelerating-expansion theory/model explains that, galaxy clusters, filled with relativistic-proton dark matter, are known to be separating from each other with velocities proportional to their separations. At the same time, when galaxy clusters' relativistic-proton dark matter erodes relativistic mass via synchrotron emission of extreme ultraviolet (EUV) or soft X-ray photons, the separation velocities between galaxy clusters will increase owing to a reduction in the gravitational tidal attraction between them.

This increases cluster-to-cluster separation that further lowers the gravitational tidal attraction between clusters, thereby further accelerating the galaxy cluster separation and thus the expansion of the universe. The force causing this accelerating separation of galaxy clusters is essentially only between galaxy clusters, not between galaxies, as will be explained.

After analysis and interpretation of astronomical data, Drexler concluded in early 2002 that dark matter is comprised primarily of relativistic protons, with the average relativistic mass of the protons being considerably higher than the mass of a proton at rest.

Astronomical data taken near Earth indicate that a hundred relativistic protons are accompanied by about 10 relativistic helium nuclei. Galactic and extragalactic magnetic fields cause most of these charged dark matter particles to remain within galaxy clusters and to emit radiant energy in the form of extreme ultraviolet (EUV), soft X-ray, or infrared photons when these charged particles cross magnetic-field lines.

This is called synchrotron emission, which is photon emission from a proton in its direction of motion. Since protons within a galaxy cluster are orbiting groups of galaxies, such photon emission from a galaxy cluster should be relatively isotropic with respect to the galaxy cluster's linear motion. Such dark matter relativistic protons within galaxy clusters would be emitting relatively high power synchrotron emission, causing them to lose kinetic energy and relativistic mass at significant rate, as if their mass were eroding.

One of the key astronomical data clues to dark energy is that dark energy pushes galaxy clusters apart, but does not push galaxies apart. How can synchrotron emission push galaxy clusters apart without pushing galaxies apart? First of all, the power of synchrotron emission from a relativistic proton in a magnetic field is directly proportional to the square of the proton's energy. Secondly, the kinetic energies estimated for the dark matter protons orbiting groups of galaxies within a typical galaxy cluster would be about 30 times higher than the energies calculated for the dark matter protons orbiting a single typical spiral galaxy, such as the Milky Way.

Thus, the synchrotron emission power per proton and relativistic mass loss rate per proton are about 900 times greater for relativistic protons orbiting groups of galaxies within a cluster compared to those protons orbiting only a single galaxy.

Therefore, the tidal gravitational force pushing galaxy clusters apart could be much greater than the tidal gravitational force attempting to push two galaxies apart.

Another question that is raised is why the accelerating expansion of the universe did not begin until about five billion years ago, as reported by astronomers. Since the universe was smaller in the prior time period, probably the smaller separations of galaxy clusters combined with the inverse-square law of gravity may have led to a higher gravitational tidal attraction between clusters that overwhelmed the relatively constant smaller synchrotron-emission repulsive effect. This logic is based the fact that in the past 13.4 billion years galaxy-cluster separation distances have increased by a factor of about 1000.

CHAPTER 20

Drexler's Dark Matter May Have Enabled First-Generation Stars to Ignite Hydrogen Fusion without Molecular Hydrogen or Dust

Aug. 5, 2009 — On July 14, Innovations-report issued a news release entitled, "Research sheds light on early star formation" [58]. The first-generation stars are referred to as primordial or Population III stars, which are very large and are comprised essentially of only hydrogen and helium, because elements such as carbon, oxygen, and nitrogen (astronomers' "metals") had not yet been created by an earlier generation of stars.

The subject of the news release was a July 9 scientific paper entitled, "The Formation of Population III Binaries from Cosmological Initial Conditions" [59] in Sciencexpress. This primordial star-formation research effort conducted at Michigan State University and Stanford University involved a cosmological computer simulation of star formation including their physical shape, but does not explain the atomic or molecular nature of the hydrogen/helium related gas constituents or how hydrogen fusion was ignited.

Apparently, there is no published generally accepted scientific explanation of how the gravitational-collapse of hydrogen and helium gas leads to a star's ignition and hydrogen fusion reaction. This subject will be explored here.

Several years ago, Bell Labs-trained discoverer of relativistic-proton dark matter, Jerome Drexler, conducted research on the formation of the first-generation stars from a new perspective. It was based upon the existence of "relativistic-proton dark matter" bombarding photons, hydrogen and helium atoms and muonic molecular ions. His scientifically plausible findings, which may provide insights into the validity and significance of the Michigan / Stanford paper, are presented as follows.

The European Southern Observatory (ESO) published an article about star formation on November 18, 2004, entitled, "Stellar Clusters Forming in the Blue Dwarf Galaxy NGC 5253" [60]. The following two excerpts from the ESO article provide some insight into the level of challenge presented to someone seeking an explanation for primordial (first created) star formation:

"Star formation begins with the collapse of the densest parts of interstellar clouds, regions that are characterized by comparatively high concentration of molecular gas and dust like the Orion complex and the Galactic Centre region. Since this gas and dust are products of earlier star formation, there must have been an early epoch when they did not yet exist."

The next paragraph continues:

"But how did the first stars then form? Indeed to describe and explain 'primordial star formation' without molecular gas and dust is a major challenge in modern astrophysics." However, Drexler's 2006 dark matter cosmology research seems to provide some possible answers.

There are at least three plausible methods for facilitating primordial star formation utilizing only atomic hydrogen and helium and the "relativistic proton dark matter" theory/cosmology. Let us begin with the simplest method.

Molecular hydrogen is needed in a galaxy disk to facilitate star formation hydrogen fusion. It is known that a mixture of 50 percent hydrogen ions and 50 percent hydrogen atoms will form molecular hydrogen much faster than would hydrogen atoms alone. Therefore, the bombardment/ionization of a galaxy's atomic hydrogen gas by its relativistic-proton dark matter, orbiting halo should facilitate the formation of hydrogen molecules and hydrogen fusion in the galaxy disk, provided that no more than 70 percent of the hydrogen is ionized.

A second plausible method for facilitating primordial star ignition, utilizing only atomic hydrogen and helium and the "relativistic-proton dark matter" theory/cosmology, involves four steps:

(1)   Relativistic-proton dark matter (and relativistic helium nuclei) colliding with compressed clouds of hydrogen and helium atoms (and photons) in a galaxy disk would generate muons that could create muonic atoms of hydrogen and of helium. (The collisions actually produce pions that decay into muons.)

(2)   In muonic atoms, the orbits of muons around protons or helium nuclei are very small because the muons weigh 207 times as much as electrons. Therefore, the positive coulomb charges of the nuclei are well shielded by the closely orbiting negative muons, making the muonic atoms have a very low effective coulomb charge, thereby enabling muonic atoms to come in close contact with one another.

(3)   Very close muonic atoms of hydrogen and helium could form muonic molecular ions with either two protons in one configuration or one proton and one helium nucleus in another configuration. These two types of molecular ions would be orbited by only one muon since the effective coulomb charge would be so low that a second muon would not bond. By this means, muonic molecular ions are formed.

(4)   Subsequent bombardment of these muonic molecular ions by the same relativistic-proton dark matter, with proton energies of about 10(15) eV and by high-energy helium nuclei, should be capable of triggering hydrogen fusion reactions and the ignition of primordial stars.

Drexler's star ignition and hydrogen fusion theory is partly based upon known astronomical data that each 10(15) eV cosmic ray proton striking Earth's atmosphere produces hundreds to one thousand muons. (The colliding cosmic ray protons [believed to be from dark matter] actually produce pions which rapidly decay into muons which, in turn, decay less rapidly into electrons, etc., in a number of microseconds. Pions are also created by non-elastic collisions between relativistic-proton dark matter and photons.)

For a number of decades, muons have been known to catalyze hydrogen fusion reactions by first forming muonic molecular ions comprised of a proton plus a helium nucleus or deuterium or another proton orbited by a muon [61][62]. Muons are also known to catalyze multiple fusion reactions since they are not destroyed in the nuclear fusion process. A Google search for "muonic hydrogen fusion" leads to dozens of Web site references.

Star-related hydrogen fusion could happen in the early universe since relativistic dark matter protons were probably multitudinous and had energies much more than a thousand times higher than what can be achieved with man-made accelerators today. Also, catalytic muons were being produced in enormous quantities, and hydrogen/helium muonic molecular ions evolved as collision targets for the dark matter relativistic protons (and relativistic helium nuclei) spiraling into galaxies from their dark matter halos.

A third plausible method for facilitating primordial star formation via relativistic-proton dark matter is based upon the theory of the starburst galaxy phenomenon. It is known that the vast majority of starburst galaxies involve the merging of two spiral galaxy clusters. According to the relativistic-proton dark matter theory/cosmology, merging of two protogalaxy (a cloud of hydrogen and helium gas forming into a galaxy) clusters would subject the hydrogen and helium in the protogalaxy disks to bombardment by relativistic protons that are orbiting groups of protogalaxies within protogalaxy clusters having proton energies about 30 times higher than dark matter protons orbiting a single protogalaxy. Such a 30-times-higher-energy proton bombardment should create higher quantities of muonic molecular ions, which then would be subjected to the same higher-energy proton bombardment, thereby facilitating hydrogen fusion ignition of the primordial stars, which essentially would be comprised of only hydrogen and helium.

Thus, three different methods of facilitating the creation of primordial stars have been presented. Depending upon conditions throughout the early universe, any one of them could be favored in some space and at some time and possibly all three methods were involved in the early epoch.

The relativistic protons in dark matter halos have sufficient energy to trigger hydrogen fusion in galaxies, as will be explained. The Web site of the Princeton Plasma Physics Laboratory-Tokomak Fusion Test Reactor (http://www.pppl.gov/projects/pages/tftr.html) reports that by 1997 the Tokamak Fusion Test Reactor [for fusion of hydrogen isotopes] achieved a world record "plasma temperature of 510 million degrees centigrade — the highest ever produced in a laboratory, and well beyond the 100 million degrees required for commercial [hydrogen] fusion." Note that in a 510 million degree plasma capable of hydrogen fusion, the average kinetic energy of the plasma protons would be far below the kinetic energy of most of the dark matter relativistic protons orbiting even a single spiral galaxy.

The catalytic capability of muons in hydrogen fusion nuclear reactions has been known for about fifty years. In 1956, at the Los Alamos Meson Physics Facility and at UC Berkeley, Luis W. Alvarez and H. Bradner discovered the hydrogen-fusion-catalytic capability of the mu-meson, now called the muon, with the help of Edward Teller. They discovered that incoming muons were able to catalyze nuclear fusion between a proton and a deuterium nucleus (one proton and one neutron). Apparently, the muons were aiding the two types of nuclei to come close enough together for quantum tunneling to allow them to fuse, "even at room temperature."

This muon-based nuclear fusion process was never commercialized because the proton bombardment energy required to produce the necessary muons was so great and the helium produced in the reaction captured so many muons (a principal source of energy loss) that the nuclear fusion process was very inefficient and impractical. However, this same or related process may be practical for creating hydrogen fusion in stars because of the extremely high energies of the multitudinous relativistic cosmic ray protons available in the universe to bombard the muonic molecular ions to trigger fusion and also to generate muons via non-elastic collisions of the protons with photons, hydrogen, and helium. This high muon creation rate could overcome and negate the muon-absorption-by-helium energy loss mentioned earlier.

Star formation does not seem to be explained satisfactorily by the decades-old generally accepted mainstream theory of star formation, where clouds of hydrogen molecules collapse anywhere in a galaxy under their own weight and are heated through compression to hydrogen fusion temperatures. Drexler encourages astrophysicists to seek star formation understanding based upon muons and upon bombardments by dark matter protons having energies at least 1000 times greater than those produce by manmade proton accelerators.

CHAPTER 21

Drexler's Relativistic-Proton Dark Matter, Not Dark Energy, Led Post-Big Bang Rapid Inflationary Epoch; BigBOSS to Test This

Aug. 13, 2009 — On August 6, 2009 an article was published in the ScienceDaily entitled, "Dark Energy From The Ground Up: Make Way For BigBOSS"[63]. The article describes a planned ground-based $71 million telescope-spectrograph system to be developed under the sponsorship of the U.S. Department of Energy. It should be capable of analyzing the mysterious dark energy responsible for the accelerating expansion of the universe, first detected in 1998, and the post-big bang inflationary epoch of the very rapid expansion of the universe, called Cosmic Inflation.

The ScienceDaily article, is based upon a 13 May 2009 revision of the scientific paper, "BigBOSS: The Ground-Based Stage IV Dark Energy Experiment"[64]. The ScienceDaily article contains the paragraph quoted here that describes a major puzzle to be solved, which triggered this newswire article response. "But BigBOSS offers more. One of the most interesting questions in cosmology is the relationship between dark energy and the early inflationary epoch of rapid expansion. Something was happening then [just after the big bang], and we wonder if it's repeating in some way. BigBOSS will have the best sensitivity to the inflationary epoch. In some ways this could be the best argument for BigBOSS of them all."

In response, Bell Labs-trained discoverer of relativistic-proton dark matter, Jerome Drexler, hereby announces that his March 2008 astro-cosmology book, *Discovering Postmodern Cosmology,* provides strong supporting evidence that the universe's post-big bang rapid inflationary epoch phenomenon, commonly called Cosmic Inflation, is unrelated to the dark energy phenomenon that has been causing the accelerating expansion of the universe for the past five to six billion years.

The nature of the dark energy/accelerating universe phenomenon is precisely explained in Chapter F, based upon relativistic-proton dark matter, without making any additional assumptions. The title of the chapter is "Eroding High-Energy Dark Matter Particles in Galaxy Clusters May Explain the Universe's Acceleration". The dark energy/accelerating universe model is based upon all galaxy clusters losing relativistic mass via synchrotron emission from the relativistic-proton dark matter racing around groups of galaxies in a cluster and through the extragalactic magnetic field. This would cause the separation velocities between all galaxy clusters to continue to increase, which would represent an accelerating expansion of the universe.

The nature of the post-big bang inflationary epoch phenomenon of the very rapid expansion of the universe, commonly called Cosmic Inflation, is described in Chapter H whose first sentence reads "How inflation happened a split second after the big bang.".

Briefly, the theory proceeds as follows: The big bang would have relativistic protons and helium nuclei being fired out at near the speed of light in many purely radial outward directions. This short first phase would be followed automatically by a second phase during which time the relativistic protons and helium nuclei would be deflected by magnetic fields and electric-charge repulsion into a transverse motion thereby greatly reducing their radial outward particle velocities and bringing to an end the universe's very rapid inflationary epoch.

The Cosmic Inflation period would be related primarily to the short first phase of almost purely radial outward particle motions near the speed of light of the relativistic protons and helium nuclei, beginning a split-second after the big bang ejected them and ending naturally as described in the previous paragraph.

The prerequisite for understanding the Cosmic Inflation theory of Chapter H is Chapter C entitled, "A Relativistic Proton Dark Matter Would Be Evidence the Big Bang Probably Satisfied the Second Law of Thermodynamics". It provides the necessary logical and plausible background for explaining Cosmic Inflation.

The paradigm-shifted cosmological conclusions arrived at in Chapters C, F, and H are based upon the laws of physics, known astronomical data, and only one assumption, that the dark matter of the universe is comprised primarily of relativistic protons (To be precise, the protons are accompanied by relativistic helium nuclei in a ratio of about one helium nucleus for every ten to twelve protons). Therefore to provide support for the claim in the title of this chapter it would be appropriate at this point to provide substantial evidence that relativistic-proton dark matter is indeed the dark matter of the universe.

The first source of evidence that relativistic-proton dark matter is indeed the dark matter of the universe proceeds as follows: No other dark matter candidate can be used to logically and plausibly explain any the following 12 cosmic phenomena or cosmic constituents known to exist in the universe without evoking additional assumptions.

However, Drexler's relativistic-proton dark matter provides plausible explanations for each and every one of them without added assumptions, as presented in his three books, two scientific papers, and/or two dozen 2008/2009 scientific articles. See the Web site:http://www.jeromedrexler.org/ .

Cosmic phenomena or cosmic constituents known to exist:

1. The accelerating expansion of the universe

2. Dark Energy

3. A dark matter that can exist in the form of spheroidal halos around spiral disc galaxies and also in the form of long large slightly curved filaments that form the Cosmic Web.

4. The source of ultra-high-energy cosmic-ray protons

5. How Cosmic Inflation started and then stopped in the post-big bang period

6. The source of the magnetic field that pervades the universe

7. Why most large galaxies formed without galaxy mergers

8. The cause for the early rapid growth of massive galaxies

9. The cause for the stunted mass growth of galaxy clusters

10. How the first stars formed without availability of hydrogen molecules or dust

11. The basis for the formation of the Lyman Alpha blobs

12. Limitation of the maximum diameter of galaxy superclusters to 430 million light years.

Remarkably, Drexler's relativistic-proton dark matter cosmology explains these 12 cosmic phenomena or cosmic constituents by utilizing only the laws of physics and known astronomical observations, without employing other assumptions. This implies that some form of relationship exists between each of these dozen cosmic phenomena or constituents and *relativistic-proton dark matter*, which makes this relativistic dark matter candidate unique.

The second source of evidence that relativistic-proton dark matter is indeed the dark matter of the universe evolves as follows: Let us start with the premise that the dark matter of the universe is comprised primarily of relativistic protons. Let us then use this premise, existing astronomical data, and the laws of physics to explain five major cosmic mysteries in a plausible manner. By accomplishing this objective five times, the premise itself becomes plausible.

This procedure is explained as follows:

The only plausible explanation for the accelerating expansion of the universe, announced to date, requires that the dark matter of the universe be comprised of relativistic protons *(see Chapter F)*.

The only plausible explanation for the big bang satisfying the Second Law of Thermodynamics, announced to date, requires that dark matter be comprised of relativistic protons *(see Chapter C)*.

The only announced plausible explanation for ultra-high-energy cosmic ray protons with energies above 60 EeV bombarding Earth's atmosphere requires that dark matter be comprised of ultra-high-energy relativistic protons *(see Chapter G)*.

The only announced plausible explanation for the Cosmic Inflation epoch requires that dark matter was comprised of ultra-high-energy relativistic protons at that time *(see Chapter H)*.

The only announced plausible explanation for the Cosmic Web requires that dark matter be comprised of relativistic protons *(see Chapter J)*.

If these five explanations are indeed plausible, it is highly probable that the dark matter of the universe has been comprised primarily of relativistic protons, reaching ultra-high-energy levels in many cases.

The third source of evidence that relativistic-proton dark matter is indeed the dark matter of the universe is as follows: An article, entitled "The Search for Dark Matter" [49], in the recent Majestic Universe special issue of Scientific American, makes the following statements about the dark matter of the universe: "(1) What kind of particle could dark matter be made of? Astronomical observation and theory provide some general clues. (2) It cannot be protons or neutrons or anything that was once made of protons or neutrons, such as massive stars that became black holes. (3) According to calculations of particle synthesis during the big bang, such particles were simply too few in number to make up the dark matter. (4) Those calculations have been corroborated by measurements of primordial hydrogen, helium, and lithium in the universe."

These four statements are true only if the protons are moving slowly, but are not true if they are relativistic protons moving near the speed of light. That is because Einstein's Special Theory of Relativity says that the relativistic mass of such very fast protons could be as much as 100, 1000, or more than a million times greater than the mass of slow-moving protons.

Now let us consider sentence (3) above, which says "such [proton or neutron] particles were simply too few in number to make up the [actual mass of] dark matter." However, from Einstein's Special Theory of Relativity we see that even if the relativistic protons are "few in number," their very high relativistic mass could still "make up the [actual mass of] dark matter." Drexler had discovered this possibility in early 2002. Drexler's confidence in the relativistic-proton dark matter model was soon enhanced when he realized that the same dark matter concept also could explain the mysterious high-energy cosmic-ray protons bombarding Earth every day that probably represent energy-reduced relativistic-proton dark matter. Their energy reduction, over billions of years, could be caused by synchrotron emission loss of photons from the relativistic protons crossing magnetic field lines in space.

Drexler became almost completely convinced of the relativistic-proton dark matter model later in 2002 when this same concept also provided a plausible explanation for the accelerating expansion of the universe, which had been discovered astronomically in 1998. Drexler's 2002 accelerating-expansion model remains today the only plausible one publicly proposed.

Drexler's explanation for it requires only basic astrophysics. It has been known through astronomy that spiral disk galaxies are enclosed by a spheroidal halo of dark matter with a mass about ten times greater than the ordinary mass of the spiral galaxy. From astronomy, we also know that a weak magnetic field at a known intensity pervades the universe. It is also well known that when relativistic protons cross magnetic field lines in space they lose relativistic mass and energy through the emission of large quantities of photons through synchrotron emission.

Furthermore, it has been known through astronomer Edwin Hubble for many years that all galaxy clusters are moving away from one another. Using all of this information, let us consider three galaxy clusters that are moving away from each other at some separation velocities.

Let us check these galaxy clusters a month later. The mass of each of the three galaxy clusters would be lower because of their continuous synchrotron emission loss of photons and thus the tidal gravitational attractive force between pairs of them also would be lower.

Therefore, their separation velocities would be higher and rising. For this same reason, all the separation velocities between galaxy clusters in the universe should be increasing, which describes, in fact, *an accelerating expansion of the universe!*

As mentioned above, the related scientific paper is entitled "BigBOSS: The Ground-Based Stage IV Dark Energy Experiment", and designated arXiv.org>astro-ph> arXiv:0904.0468v3 [64].

Its Abstract reads:

"The BigBOSS experiment is a proposed DOE- NSF Stage IV ground-based dark energy experiment to study baryon acoustic oscillations (BAO) and the growth of structure with an all-sky galaxy redshift survey. The project is designed to unlock the mystery of dark energy using existing ground-based facilities operated by NOAO. A new 4000-fiber R=5000 spectrograph covering a 3-degree diameter field will measure BAO and redshift space distortions in the distribution of galaxies and hydrogen gas spanning redshifts from 0.2<Z."

"This project will enable an unprecedented multi-object spectroscopic capability for the U.S. community through an existing NOAO facility. The U.S. community would have access directly to this instrument/telescope combination, as well as access to the legacy archives that will be created by the BAO key project."

CHAPTER 22

Soft X-Ray Emission from Dark Matter within Galaxy Clusters, Predicted by Drexler in 2006, Now Goal of Russian-German Team

Aug. 25, 2009 — On August 19, 2009, a news release [65] was published in Moscow by RIA Novosti entitled, "Russia, Germany to jointly probe universe for 'dark matter.'" They plan [66] to be the first to use a satellite X-ray telescope to search for dark matter. This surprising news comes three years after Bell Labs-trained discoverer of relativistic-proton dark matter, Jerome Drexler, published his July 2006 prediction that dark matter within galaxy clusters should exhibit soft X-ray photon emission via synchrotron emission, which occurs when relativistic protons cross magnetic field lines.

To date, no astronomer, astrophysicist, or cosmologist has published confirmation, support or encouragement for Drexler's prediction of soft X-ray photon emission or extreme UV photon emission from dark matter flowing within galaxy clusters. Thus the Russian-German news, about efforts to do just that, was warmly welcomed by Drexler.

Some relevant sentences from the Russian news release are as follows:

"A Russian-German orbital observatory set for launch in 2012 will help scientists to study the role of 'dark matter' in the structure and the evolution of the universe, a Russian research institute said on Wednesday."

"Russia's Federal Space Agency and the German Aerospace Center signed on Tuesday an agreement on the launch of the German eROSITA X-ray telescope and the Russian ART-XC device [for X-ray spectroscopy and time analysis of galactic and extragalactic radiation] on board the Russian Spectrum-Roentgen-Gamma (SRG) satellite."

"The observatory will use the most sensitive to date high-energy mirror detectors to survey about 100,000 galaxy clusters in the search for dark matter, which is undetectable but necessary to explain observed phenomena in the universe and currently accepted cosmological theories."

Drexler had predicted the emission of soft X-ray photons, extreme ultraviolet (EUV) photons, or UV photons from relativistic-proton dark matter orbiting a group of two or three spiral galaxies in the following excerpt from his July 17, 2006 newswire article, which is also published in Chapter 2 of his March 2008 paperback book entitled *Discovering Postmodern Cosmology*.

"LOS ALTOS HILLS, Calif., July 17, 2006 (AScribe Newswire) — Ever since the concept of galaxy-orbiting relativistic proton dark matter was developed by Jerome Drexler over five years ago, he has been aware that his dark matter model should emit synchrotron radiation of ultraviolet photons, extreme ultraviolet (EUV) photons, or soft X-ray photons. The extragalactic magnetic field is at the right level."

"In the past, astronomers doubting Drexler's relativistic proton dark matter model had argued that if his model were correct, ultraviolet or EUV photons emitted from dark matter halos around spiral galaxies would have been detected by astronomers years ago."

"Drexler had responded that the failure of astronomers to detect the synchrotron radiation probably was caused by the low power level of the emitted ultraviolet photons that are generated by relativistic protons moving near the speed of light in circular orbits that form dark matter halos around [individual] spiral galaxies. However, [relativistic] protons orbiting clusters of [two or three of] these spiral galaxies are much more energetic and could generate photons with hundreds of times more synchrotron radiation power in the extreme ultraviolet (EUV) [or soft X-rays, as in the above prediction]."

Drexler also has pointed out that the absorption by Earth's atmosphere is very effective in filtering out ultraviolet, extreme ultraviolet, and soft X-ray photons. Thus, to detect UV or X-rays from relativistic-proton dark matter would require a satellite telescope observatory. Therefore, NASA's installation of the UV Cosmic Origins Spectrograph (operational wavelengths from 115 to 320 nanometers) on the Hubble Space Telescope on May 16, 2009 may turn out to be a dream come true for Drexler.

Furthermore, Russia earlier announced that it will launch the Spektr-UF ultraviolet astronomical observatory, which would have a 1.7 meter diameter main mirror, into a highly elliptical orbit in 2010. In the news release, the Russian project manager and Professor of Physics and Mathematics Boris Shustov, who heads the Institute of Astronomy of the Russian Academy of Sciences, is quoted as saying, "One should particularly emphasize the observatory's role in detecting the so-called dark matter of the Universe and unlocking its secrets because such dark matter can only be seen by large ultraviolet telescopes."

This also could be a dream come true for Drexler.

Thus, one might conclude from the above that the combination of (1) Professor Boris Shustov's 2010 ultraviolet satellite telescope research program following (2) the 2009 NASA/Hubble's installation of the UV-based Cosmic Origins Spectrograph (COS) to detect dark matter in the Cosmic Web, followed by the use of (3) the Russian-German soft X-ray satellite telescope in 2012 to search for dark matter in galaxy clusters may confirm Drexler's relativistic-proton dark matter as the universe's dark matter.

Drexler has documented his seven years of dark matter/dark energy research in the following list of six publications:

(1) Paperback book, November 2009, *Our Universe via Drexler Dark Matter: Drexler Dark Matter Created &Explains Dark Energy, Top-Down Cosmology, Inflation, Accelerating Cosmos, Stars, Galaxies, Cosmic Web.*

(2) Paperback book, March 1, 2008, *Discovering Postmodern Cosmology: Discoveries in Dark Matter, Cosmic Web, Big Bang, Inflation, Cosmic Rays, Dark Energy, Accelerating Cosmos.*

(3) Scientific paper, physics/0702132, Feb. 15 2007, "A Relativistic-Proton Dark Matter Would Be Evidence the Big Bang Probably Satisfied the Second Law of Thermodynamics."

(4) Paperback book, May 22, 2006, *Comprehending and Decoding the Cosmos: Discovering Solutions to Over a Dozen Cosmic Mysteries by Utilizing Dark Matter Relationism, Cosmology, and Astrophysics.*

(5) Scientific paper, astro-ph/0504512, April 22, 2005, "Identifying Dark Matter through the Constraints Imposed by Fourteen Astronomically Based 'Cosmic Constituents'".

(6) Paperback book, Dec. 15, 2003, *How Dark Matter Created Dark Energy and the Sun: An Astrophysics Detective Story.*

APPENDIX I

PREREQUISITE CHAPTER A

Overview of Search for Solutions in Dark Matter Cosmology and Postmodern Cosmology

February 11, 2008 — The author believes that the galaxy-orbiting relativistic proton appears to have the necessary characteristics of the long-sought dark matter particles, which are estimated by most scientists to comprise about 83 percent of the total mass of the universe. Relativistic protons do have the required mass and the required difficulty of detection and can transform themselves into hydrogen, the principal matter of galaxies, by combining with electrons created through pion producing collisions and pion decay.

Therefore, relativistic protons are capable of forming (1) galaxies and their dark matter halos, (2) galaxy clusters and their distributed dark matter, (3) the Cosmic Web[33,43,77], the long, large, filamentary dark matter that crisscrosses the cosmos, and (4) newborn stars and igniting their hydrogen fusion reactions.

However, for this relativistic-proton dark matter theory to become widely accepted, there also should be astronomical evidence of multitudinous relativistic protons within the spheroidal dark matter halo surrounding the Milky Way and other spiral galaxies. The author believes that the energetic cosmic ray relativistic protons bombarding Earth every day go a long way toward providing such astronomical evidence.

A vast majority of Cosmologists have not yet accepted the author's explanation of the nature of dark matter, the Cosmic Web, dark energy, the big bang, the accelerating expansion of the universe, the energy source(s) for ultra-high-energy cosmic ray protons, or cosmic inflation. Hopefully, this book will provide sufficient scientific evidence to convince some of the cosmologists of the validity or plausibility of some of these explanations.

In order to strengthen his case for relativistic-proton dark matter with the cosmologists, the author devised in 2005 a second independent approach, beyond his 2003 published approach, to confirm the identity of dark matter. Since dark matter represents about 83 percent of the mass of the universe, it is omnipresent and should have an influence on or a relationship with a number of celestial bodies. Such relationships might be used to identify dark matter, he felt.

Meanwhile, the vast majority of research conducted on dark matter by physicists has had to do with trying to identify the particles that comprise dark matter or to determine their gravitational effect on star rotation curves in spiral galaxies. This primarily inward-looking approach to identify the particle composition of a medium is known as reductionism, *which is a procedure or theory that reduces or attempts to reduce complex data or phenomena into simple elements.*

Reductionism does not always work well in physics. Many times simple entities or particles can form complex forms or combinations that have characteristics seemingly unrelated to the characteristics of the original simple entities. A hurricane is one well-known example of complex behavior whose characteristics cannot be predicted by an analysis of all the known simple entities involved in its makeup. Thus, the reductionism approach does not explain or predict the nature of a hurricane.

An alternative to the inward-looking reductionism is an outward-looking, cosmological-like approach that the author has developed and designated *relationism,* where a phenomenon such as the dark matter can be analyzed and categorized in terms of its various relationships. *Dark matter relationism,* described in the author's 2006 book, is used to provide additional evidence that relativistic-proton dark matter is the dark matter of the universe.

The 2008 book provides a third independent approach to build the case for relativistic-proton dark matter. It argues that the big bang, which occurred at the beginning of time, must have satisfied the Second Law of Thermodynamics. Thus, immediately after the extremely high energy big bang, the entropy (disorder) of the universe would be at the lowest level it would reach throughout all time. This could be achieved by the big bang firing out, in all directions, high-velocity ultra-high-energy (UHE) relativistic protons and helium nuclei in the well-known nuclei ratio near 12:1.

Such a big bang, characterized by a *violent radial dispersion* of ultra-high-energy relativistic nuclei, would be highly efficient and could create very high usable energy and have very low entropy, and might be designated a *Relativistic Big Bang*. This big bang concept is fundamental to the author's relativistic-proton dark matter theory and to his postmodern big bang cosmology.

Postmodern cosmology and postmodern big bang cosmology mean the same thing. Their explanation begins with requiring that the big bang satisfy the Second Law of Thermodynamics, which essentially requires that the vast majority of mass/energy produced by the big bang be in the form of relativistic protons and helium nuclei. Dark matter represents 83 percent of the mass of the universe, and was produced by the big bang. Therefore dark matter would have to be comprised of relativistic protons and helium nuclei.

Over 99 percent of the mass of the universe is hydrogen and helium in the atomic ratio of about 12:1. The big bang created all the mass of the universe, with almost all consisting of relativistic protons and helium nuclei in the atomic ratio not far from 12:1. Since 83 percent of the mass of the universe is dark matter, it follows that dark matter should be comprised of relativistic protons and helium nuclei in a ratio not far from 12:1. The name relativistic-proton dark matter is defined by this specific mix of protons and helium nuclei, which are both baryons. In 2008, the more accurate name, relativistic-baryon dark matter began to be used.

In this book we will develop the only publicly announced plausible explanations for *dark matter*, the *Cosmic Web*, the *big bang*, the *accelerating expansion of the universe*, for the existence of the *ultra-high-energy cosmic-ray protons*, for *cosmic inflation*, and insights into the nature of *dark energy*.

Relativistic-proton dark matter is also used to explain, in the author's 2006 book, how the first stars were created, why a spiral galaxy creates blue stars in its spiral arms, why starburst galaxies create blue stars in their cores and have star formation rates fifty times higher than that of a spiral galaxy, and why there is a different form of dark matter around small galaxies as compared to dark matter around groups of galaxies within normal galaxy clusters.

Postmodern cosmology requires a change of focus, for physicists and cosmologists, onto relationism rather than only on reductionism. It requires that the universe be looked upon as a complex, orderly, self-inclusive system in space and time comprised of interrelated cosmic entities, rather than simply a collection of unrelated cosmic entities.

A number of the chapters provide published research results from other scientists casting doubt on the existence of Cold Dark Matter , which remains intangible 23 years after it was proposed. In contrast, the author is not aware of published scientific research data casting doubt on the existence of the 4-year-old relativistic-baryon dark matter.

Therefore, it seems as if a dark-matter *fork in the road* has appeared on the horizon. In order for cosmologists to make significant progress on *postmodern cosmology* they will have to decide which dark-matter fork to take. Study of this book could be helpful in analyzing the alternatives.

PREREQUISITE CHAPTER B

Dark Matter's Identity Revealed
by Deciphering 14 Cosmic Clues

September 5, 2006 — Dark matter's identity has been discovered through use of a cryptographic-like analysis of 14 constituents of the cosmos.

As a youth during wartime, Jerome Drexler learned how to decipher a 50-word encrypted message or a 50-word encrypted passage from Shakespeare. A decade later in graduate school, a course in Information Theory expanded his knowledge of cryptography.

Drexler has applied a cryptographic-like analysis for solving the mystery of the identity of dark matter (DM) of the universe. Instead of using a 50-word encrypted message to extract the secret code it contains, he used 14 carefully selected cosmic clues called cosmic constituents of the universe to extract the nature and identity of dark matter.

He had speculated that if dark matter represents 80 to 90 percent of the mass of the universe, dark matter should have roles, functions and an influence on most of these 14 cosmic constituents. Each type of dark matter proposed by scientists was subjected to 14 elimination tests as follows.

Drexler asked 14 rhetorical questions: Which type of dark matter (DM) particles could:

1. Form spheroidal dark matter halos around galaxies and DM halos around galaxy clusters?

2. Cause the accelerating expansion of the universe and possibly store dark energy?

3. Be transformed into low velocity hydrogen, protons or proton cosmic rays?

4. Create the magnetic fields within and around spiral galaxies?

5. Be concentrated in the long large curved filaments of dark matter, announced by NASA[33,77] on September 8, 2004, which form galaxy clusters where two DM filaments intersect (known as the *Cosmic Web*)?

6. Create large mature spiral galaxies less than 2.5 billion years after the big bang?

7. Create spheroidal DM halos having predictable outer and *hollow* core diameters?

8. Provide angular momentum to spiral galaxies and DM halos?

9. Create galaxies without a central DM density cusp?

10. Create a starless galaxy or a LSB dwarf galaxy with low star formation rates?

11. Lead to linearly rising rotation curves for LSB dwarf galaxies and to flat rotation curves for spiral galaxies?

12. Form 80 percent to 90 percent of the mass of the universe, the remainder being hydrogen, helium, etc?

13. Ignite hydrogen fusion reactions of second-generation stars utilizing hydrogen molecules and dust and ignite fusion reactions of the first generation stars with only hydrogen atoms?

14. Create the first *knee* at 3x1015 eV, the second *knee* between 1017 eV and 1018 eV and the *ankle* at 3x1018 eV of the cosmic-ray energy spectrum near the Earth?

After careful study and analysis, Drexler concluded that galaxy-orbiting relativistic protons would provide many more affirmative answers to the 14 questions than any other known particle. Therefore relativistic-proton dark matter could be the identity of dark matter since it appears to have the strongest influence on and relationship with the 14 cosmic constituents.

Relativistic-proton dark matter satisfies the three basic requirements of a dark matter candidate. Do such protons have sufficient mass? Yes, relativistic protons can have enormous mass. Have they ever been detected? Yes, relativistic protons bombard Earth's atmosphere every day and are called cosmic rays. Don't relativistic protons move too fast to form small galaxies? The protons can form small galaxies after the protons are slowed down by muon-producing collisions and synchrotron emission losses and after the protons combine with the electrons created by the muon decay, thereby forming hydrogen.

Since protons are electrically charged particles, they would be constrained by the galactic and extragalactic magnetic fields into circular-type orbits forming dark matter halos around galaxies and dark matter around groups of galaxies within galaxy clusters, and also would be concentrated in long large curved filaments of dark matter[33]. All three of these dark matter configurations have been detected by astronomers.

Drexler's research has led not only to the identification of the dark matter, but also to the discovery of the surprising and significant roles and functions of dark matter in creating spiral galaxies, stars, starburst galaxies, extreme ultraviolet synchrotron radiation, and the ultra-high-energy cosmic rays that bombard Earth. Dark matter appears to be a very active and dynamic medium, not the passive medium represented by Cold Dark Matter.

PREREQUISITE CHAPTER C

Relativistic-Proton Dark Matter Would Be Evidence the Big Bang Probably Satisfied the Second Law of Thermodynamics*

Abstract

February 15, 2007 A new research hypothesis has been developed by the author based upon finding astronomically based *cosmic constituents* of the universe that may be created or influenced by or have a special relationship with possible dark matter (DM) candidates. He then developed a list of 14 relevant and plausible *cosmic constituents* of the Universe, which then was used to establish a list of constraints regarding the nature and characteristics of the long-sought dark matter particles. A dark matter candidate was then found that best conformed to the 14 constraints established by the *cosmic constituents*. The author then used this same dark matter candidate to provide evidence that the big bang could be characterized as a *violent radial dispersion of relativistic baryons*, had a low entropy, and therefore probably satisfied the Second Law of Thermodynamics.

Determining the Nature of the
Dark Matter of the Universe

One hundred years ago, Albert Einstein announced the Special Theory of Relativity, which predicted and explained that a proton traveling near the speed of light could have a relativistic mass a thousand, a million, or even a billion times greater than the mass of a proton at rest. (This led the author to conceive his dark matter theory. The idea occurred to him that the gravitational strength of multitudinous galaxy-orbiting relativistic protons moving in the cosmos could create extremely large gravity-related tidal forces on nearby matter, like that exhibited by dark matter.)

Astronomer Fritz Zwicky [45] discovered the presence of dark matter in the Coma cluster of galaxies in 1933. Ever since astronomer Vera Rubin [67,68] confirmed the existence of dark matter halos around galaxies in 1977, cosmologists and astrophysicists have been trying to identify the dark matter particles.

In 1984, scientists[1] developed a Cold Dark Matter (CDM) theory based upon a theoretical uncharged, slow moving particle that they called the Weakly Interacting Massive Particle (WIMP). More recently, it was estimated by scientists that the theoretical WIMP dark matter particles would require a mass in the range of about 35 to 10,000 times [34] greater than the mass of a proton at rest in order to exhibit the observed gravity-related forces of dark matter halos. However, searches for the theoretical WIMP particles during the past 20 years have all come up empty handed.

For this reason, and knowing that Einstein's relativistic proton easily could meet the mass requirement of the mysterious dark matter particles and that relativistic cosmic ray protons are widely observed, the author has endeavored to determine the nature of dark matter.

The author posits that relativistic protons, orbiting galaxies, have the necessary characteristics of the long-sought dark matter particles, which are estimated by most scientists to comprise 80% to 90% of the total mass of the universe. Relativistic protons do have the required mass and the required difficulty of detection. Protons also can transform themselves into hydrogen, the principal matter of galaxies, by creating muons[61,35] that decay into electrons, then combining with the electrons.

Thus, relativistic protons could form (1) galaxies and their dark matter halos, (2) galaxy clusters and their internal dark matter, and (3) the long, large, filamentary dark matter known [33,43,77] to crisscross the cosmos [now called the Cosmic Web].

However, for this proton-based dark matter theory to become widely accepted, there also should be astronomical evidence of relativistic protons within the dark matter halo surrounding the Milky Way. The author posits that the high-energy cosmic ray relativistic protons bombarding Earth's atmosphere every day, from all directions, lend credence toward providing such astronomical evidence.

The author has applied a cryptographic-like analysis for solving the mystery of the identity of dark matter of the universe. Instead of using an encrypted message to extract the secret code it contains as in normal cryptography, the author used 14 cosmic constituents of the universe to extract the nature and identity of dark matter.

The author had speculated that if dark matter represents 80% to 90% of the mass of the universe, dark matter should have roles, functions or an influence on most of the following 14 cosmic constituents. Each type of dark matter proposed by scientists was subjected to 14 elimination tests as follows.

The author asked 14 rhetorical questions: Which type of dark matter (DM) particles could:

1. Form spheroidal dark matter halos around galaxies and DM halos around galaxy clusters?
2. Cause the accelerating expansion of the universe and possibly store dark energy?
3. Be transformed into low-velocity hydrogen, protons, or proton cosmic rays?
4. Create the magnetic fields within and around spiral galaxies?

5. Be concentrated in the long, large, curved filaments of dark matter announced by NASA on September 8, 2004[33,43] [now called the *cosmic web*], which form galaxy clusters where two DM filaments intersect?

6. Create large, mature, spiral galaxies less than 2.5 billion years after the big bang?

7. Create spheroidal DM halos having predictable outer and *hollow* core diameters?

8. Provide angular momentum to spiral galaxies and DM halos?

9. Create galaxies without a central DM density cusp?

10. Create a starless galaxy or a Low Surface Brightness (LSB) dwarf galaxy with low star formation rates?

11. Lead to linearly rising rotation curves for LSB dwarf galaxies and to flat rotation curves for spiral galaxies?

12. Form 80% to 90% of the mass of the universe, the remainder being hydrogen, helium, etc.?

13. Ignite hydrogen fusion reactions of second generation stars utilizing hydrogen molecules and dust and ignite fusion reactions of the first generation stars with only hydrogen atoms?

14. Create the first *knee* at 3×10^{15} eV, the second *knee* between 10^{17} eV and 10^{18} eV, and the *ankle* at 3×10^{18} eV of the cosmic-ray energy distribution at the Earth? *(See Appendix II, Slide #17)*

After careful study and analysis, the author concluded that galaxy-orbiting relativistic protons would provide many more affirmative answers to the 14 questions than any other known particle. Therefore, relativistic-proton dark matter could be the identity of dark matter since it appears to have the strongest influence on and relationship with the 14 *cosmic constituents*. This dark matter identification procedure could also be described as utilizing Ockham's (Occam's) razor logic 14 times.

Relativistic-proton dark matter satisfies the three basic requirements of a dark matter candidate. Do such protons have sufficient mass? Yes, relativistic protons can have enormous mass. Have they ever been detected? Yes, relativistic protons bombard Earth's atmosphere every day and are called cosmic rays. Don't relativistic protons move too fast to form small galaxies? The protons can form small galaxies after the protons are slowed down by muon-producing [61,35] collisions and synchrotron emission energy losses, and after the protons combine with the electrons created by the muon decay, thereby forming hydrogen.

Since protons are electrically charged particles, they would be constrained by the weak extragalactic and galactic magnetic fields into extremely large circular/spiral orbits forming dark matter halos around galaxies and dark matter around groups of galaxies within galaxy clusters, and also could be concentrated into long large curved filaments of dark matter. All three of these dark matter configurations have been reported by astronomers.

Much of the above information was derived from the author's May 2006 book [35] and his 19-page April 2005 paper, "Identifying Dark Matter through the Constraints Imposed by Fourteen Astronomically Based Cosmic Constituents"[15], found on the arXiv.gov website as e-print No. astro-ph/0504512.

The author's 295-page May 2006 book, *Comprehending and Decoding the Cosmos*, analyzes an additional 11 cosmic enigmas beyond the 14 derived from his astro-ph paper [15]. Utilizing only Relativistic-Proton Dark Matter theory and the laws of physics, the author explains in a plausible manner all 11 of these recently discovered cosmic enigmas, further supporting the relativistic-proton dark matter theory.

The author's research has led not only to the identification of the dark matter but also to the discovery of the surprising and significant roles and functions of dark matter in creating the Cosmic Web, spiral galaxies, stars, starburst galaxies, extreme ultraviolet synchrotron radiation, and the ultra-high-energy cosmic rays that bombard Earth.

Dark matter appears to be a very active and dynamic medium comprising relativistic protons and helium nuclei in the well-known ratio of about 12: 1. Dark matter is widely believed to represent 80% to 90% of the mass of the universe, and believed to be created by the big bang. These dark matter characteristics provide the evidence required and used in the next section to reach the conclusion that the big bang was relativistic, had a low entropy, and probably satisfied the Second Law of Thermodynamics.

A Relativistic-Proton Dark Matter Would Be Evidence that the Big Bang Had Low Entropy and Probably Satisfied the Second Law of Thermodynamics

In a surprising manner, the big bang may have satisfied the Second Law of Thermodynamics. An understanding of this phenomenon is helped by an excerpt from Stephen Hawking's earlier tutorial [69] on the subjects of disorder, entropy, the Second Law of Thermodynamics, and the arrow of time: "It is a matter of common experience that things get more disordered and chaotic with time. This observation can be elevated to the status of a law, the so-called Second Law of Thermodynamics. This says that the total amount of disorder, or entropy, in the universe, always increases with time."

If the amount of disorder, or entropy, in the universe always increases with time, then at the beginning of time the entropy must have been at its lowest level. The big bang also occurred at the beginning of time. Therefore, if we accept the Second Law of Thermodynamics, we must also accept that immediately after the big bang the entropy of the universe would be at the lowest level it would reach throughout all time.

However, the big bang is normally characterized as a fiery, chaotic, fireball explosion associated with a high level of disorder and entropy. We are thus faced with an enigma as to the level of entropy following the big bang, but we are not alone.

On November 18, 2004, the University of Chicago published an article [70] entitled "Astrophysicists attempt to answer the mystery of entropy" that contains the following relevant two-sentence paragraph: "But the mystery remains as to why entropy was low in the universe to begin with. The difficulty of that question has long bothered scientists, who most often simply leave it as a puzzle to answer in the future."

If the entropy following the big bang had been very low, the Second Law of Thermodynamics would have been satisfied, but how could a fiery, chaotic big bang explosion have low entropy? This is the enigma that "has long bothered scientists."

The author sees a possible solution to this enigma that would have the big bang firing out, in all directions, high-speed ultra-high-energy (UHE) relativistic protons and helium nuclei near the well-known atomic ratio of 12:1; In other words, a *violent radial dispersion of relativistic baryons.*

A very high percentage of their energies would be available to do work in the universe while their entropy, the measure of the amount of their energy which is unavailable to do work, would be very low. Such a big bang, characterized by a *violent radial dispersion* of UHE relativistic nuclei, could create very high usable energy and very low entropy, and could be designated a *Relativistic Big Bang (RBB)*.

The temperature of a *Relativistic Big Bang* could be estimated by averaging the kinetic energies of the relativistic protons and helium nuclei. The estimated temperature probably would be of the same order of magnitude as the temperature that scientists estimate for the big bang. Nevertheless, the *Relativistic Big Bang* would have the very low entropy that the Second Law of Thermodynamics requires for the *beginning of time*.

Some astronomical evidence for a *Relativistic Big Bang* comes from the ultra-high-energy cosmic ray (UHECR) protons that bombard the Earth's atmosphere every day. The *RBB* is the most plausible origin of these UHECRs (See Chapters G and H regarding the GZK effect.) In the author's relativistic-proton dark matter theory, these UHECRs are stragglers from the UHE relativistic protons that orbit groups of galaxies within galaxy clusters.

It is widely accepted that the mass of dark matter today totals about 83% of the mass of the universe and that dark matter was created by the big bang. Because of this very strong big bang-dark matter linkage, strong evidence of the existence of relativistic-proton dark matter would provide strong evidence for the existence of the *Relativistic Big Bang.* The author believes that his 2003 and 2006 books and his 2005 scientific paper provide very strong scientific evidence for the existence of relativistic-proton dark matter and, therefore, for the existence of the *RBB.*

Cosmological support for an *RBB* may come eventually via compatibility with, for example, the Cosmic Microwave Background, or Cosmic Inflation, or the Second Law of Thermodynamics, or the temperatures of the big bang, or the mass values for dark matter particles, or the 83% dark matter mass. Note that an *RBB* would be a very efficient way of creating the universe and conserving its energy because the fewest number of particles and the most useful energy would be created and dispersed, which are characteristics that may be compatible with Cosmic Inflation theory and its associated big bang.

As previously indicated, strong scientific evidence that the dark matter of the universe is comprised of galaxy-orbiting relativistic protons can be found in the 2003 book *How Dark Matter Created Dark Energy and the Sun,* the 2005, 19-page scientific paper "Identifying Dark Matter Through the Constraints Imposed by Fourteen Astronomically Based 'Cosmic Constituents'", and the 2006 book, *Comprehending and Decoding the Cosmos: Discovering Solutions to Over a Dozen Cosmic Mysteries by Utilizing Dark Matter Relationism, Cosmology, and Astrophysics* .

Confirmation of the identification of dark matter is scientifically supported in the 2006 book through the utilization of the relativistic-proton dark matter hypothesis, in conjunction with the laws of physics, to derive solutions and plausible explanations for more than 15 previously unsolved cosmic mysteries.

If the existence of the relativistic-proton dark matter provides strong evidence that the big bang satisfied the Second Law of Thermodynamics, then a corollary could follow: Since the big bang must have satisfied the Second Law of Thermodynamics, its entropy must have been very low; and since relativistic protons possess the highest possible energy and the lowest possible entropy, they must have represented the principal mass output of the big bang.

PREREQUISITE CHAPTER D

'Ring of Dark Matter' Uncovered from Anomalies/ Discrepancies

May 14, 2007 — The 'Ring of Dark Matter' was uncovered from the NASA-Hubble 3D Dark Matter Map data after astro-cosmology author Jerome Drexler pointed out on January 16, 2007 that the anomalies and discrepancies reported by astronomers on January 7 were actually valid data. NASA issued the following news advisory May 10.

"GREENBELT, Md. — NASA will hold a media teleconference at 1 p.m. EDT on May 15 to discuss the strongest evidence to date that dark matter exists. This evidence was found in a ghostly ring of dark matter in the cluster CL0024+17, discovered using NASA's Hubble Space Telescope. The ring is the first detection of dark matter with a unique structure different from the distribution of both the galaxies and the hot gas in the cluster. The discovery will be featured in the June 20 issue of the Astrophysical Journal." (A paper was also published May 2007 online as arXiv: 0705.2171[71], titled "Discovery of a Ringlike Dark Matter Structure in the Core of the Galaxy Cluster Cl0024+17".)

In early January 2007 the original research paper, "Dark matter maps reveal cosmic scaffolding," was presented and a press conference was held. The researchers stated their concerns about anomalies and discrepancies they had observed, "*naked* clumps of dark matter where there were no galaxies." Ten such statements are listed in a BBC NEWS article "Hubble makes 3D dark matter map"[74].

The researchers could not comprehend the existence of dark matter alone without galaxies. Jerome Drexler sensed that they must be relying on the unproven bottom-up theory of galaxy formation. He felt they should utilize his top-down theory of galaxy formation, in which dark matter can exist alone. With that in mind, he distributed the following Jan. 16, 2007 AScribe Newswire to the media and to 1000 dark-matter and NASA astronomers, astrophysicists, and cosmologists. Apparently it worked.

So-Called Anomalies in NASA-Hubble 3D Dark Matter Map are Explained by Astro-Cosmology Author Jerome Drexler

LOS ALTOS HILLS, Calif., Jan. 16, 2007 (AScribe Newswire) — The recent 3D mapping of the dark matter of the universe is a major astronomical accomplishment of NASA and the Hubble telescope. However, the researchers' reports of so-called discrepancies and anomalies in the distribution of dark matter relative to the distribution of ordinary matter has prevented the 3D mapping from being an immediate cosmological success.

A study of the researchers' Jan. 7 comments during a press conference following the presentation of their scientific paper at the 209th meeting of the American Astronomical Society (AAS) in Seattle, Wash., provides some insight into the reasons behind the researchers' reports of apparent discrepancies and anomalies. (See their comments below.)

Based upon their stated concerns about the so-called discrepancies and anomalies, the researchers apparently believe in the bottom-up theory of galaxy formation, not the top-down theory of galaxy formation, which has some support including that of Jerome Drexler.

The astronomical data the researchers are concerned about actually support the top-down theory of galaxy formation and if the data had been described in that manner there would not have been the issue of discrepancies or anomalies — and the 3D mapping might have been considered a cosmological success as well as an astronomical success.

Some relevant comments by some of the dark-matter-map researchers at an AAS press conference in Seattle on Jan. 7, 2007 are as follows:

> "We have to resolve discrepancies in the otherwise strong connection between ordinary matter and dark matter."

> "Finding what I would call 'naked' clumps of dark matter where there are no galaxies for me is very strange. All dark matter clumps of sufficient size should have galaxies — if our understanding is correct."

"We see that dark matter concentrations sometimes seem to have no corresponding ordinary matter."

"For the moment, no one is talking about needing to revise cosmological models; but everything hinged on the size of these anomalies."

"The discrepancies could turn out simply to be artifacts, caused by noise in the data. But then again, they could be real."

A researcher said the anomalies were *tantalizing* and that his team was eager to investigate them more closely.

The bottom-up theory, where small galaxies form first and then merge over time to form large galaxies, is known to have serious flaws. For example, note that at the same AAS conference, New York University researchers gave a paper based upon their astrophysics paper, "A New Force in the Dark Sector?"[12] Their paper states, "The number of superclusters observed in SDSS data appears to be an order of magnitude [about ten times] larger than predicted by Lambda-Cold Dark Matter simulations."

Since Lambda-Cold Dark Matter simulations are based on the bottom-up theory of galaxy formation, the NYU researchers are indicating that the bottom-up theory is extremely inaccurate in predicting the number of superclusters in the universe. Therefore the so-called anomalies and discrepancies reported by the 3D dark-matter-map researchers may have evolved simply because the researchers tested the 3D astronomical data against the wrong galaxy-formation theory. The NASA-Hubble researchers should now test the so-called discrepancies and anomalies against the top-down theory of galaxy formation.

In the top-down theory of galaxy formation the well-known long, large, slightly curved dark matter filaments form potential galaxy clusters where such dark matter filaments intersect and collide[33]. Then small and large galaxies form in this potential-cluster region from the *remnants* of the collisions of the intersecting dark matter filaments. (In the case of relativistic-proton dark matter, the *remnants* are the useful protons and helium nuclei.)

The top-down theory of galaxy formation is further explained in the pages of ten Index references in Drexler's May 2006, 295-page astro-cosmology book titled *Comprehending and Decoding the Cosmos: Discovering Solutions to Over a Dozen Cosmic Mysteries by Utilizing Dark Matter Relationism, Cosmology, and Astrophysics.*

(This ends the January 16, 2007 newswire, which was sent to the media and to 1000 dark-matter and NASA astronomers, astrophysicists, and cosmologists.)

PREREQUISITE CHAPTER E

Science Magazine's Dark Matter, 'Hydrogen in Some Hard-to-Trace Form', Opens Door to Relativistic-Proton Dark Matter

May 29, 2007 — The scientific paper, "Missing Mass in Collisional Debris from Galaxies" [8] in the May 25 issue of Science magazine is significant in that it questions the 23-year-old mainstream Cold Dark Matter (CDM) theory, and it also opens the door of scientific acceptance to the competing five-year-old relativistic-proton dark matter cosmology.

The researchers' conclusion, a significant departure from Cold Dark Matter theory, reads: "it more likely indicates that a substantial amount of dark matter resides within the disks of spiral galaxies. The most natural candidate is molecular hydrogen in some hard-to-trace form."

The researchers point out that their conclusions disagree with the Cold Dark Matter theory that posits that (1) there is no dark matter in the disks of spiral galaxies and (2) that dark matter is comprised of non-baryonic matter, which excludes hydrogen and protons.

In agreement with the researchers' conclusion is the five-year-old competing relativistic-proton dark matter theory and cosmology that posits that relativistic-protons, a hard-to-trace form of hydrogen, does indeed reside within the disks of spiral galaxies, as well as in their halos.

The Science paper clearly establishes new constraints on the nature and location of dark matter in spiral galaxies and in recycled-from-debris dwarf galaxies. The paper carefully analyzes astronomical dark matter in a triplet of recycled dwarf galaxies formed from debris from the collision of two massive spiral galaxies. The current mainstream Cold Dark Matter theory hypothesizes that such recycled-from-debris dwarf galaxies should be free of non-baryonic dark matter.

However, all three of the recycled dwarf galaxies were discovered to have twice as much dark matter as ordinary matter. Therefore the researchers were forced to conclude that the dark matter in debris-based dwarf galaxies must be baryonic since it could not be non-baryonic. They further concluded that the recycled dwarf galaxy's baryonic dark matter would have come from the disks of the colliding massive spiral galaxies.

The researchers' conclusion that the disks of spiral galaxies harbor "molecular hydrogen in some hard-to-trace-form" opens the door of scientific acceptance to the five-year-old relativistic-proton dark matter cosmology. This relatively new dark-matter cosmology is described in two recently published books and in two recent scientific papers, all authored by Silicon Valley's J.Drexler.

The abstract of the Science paper reads as follows: "Recycled dwarf galaxies can form in the collisional debris of massive galaxies. Theoretical models predict that, contrary to classical galaxies, these recycled galaxies should be free of non-baryonic dark matter. By analyzing the observed gas kinematics of such recycled galaxies with the help of a numerical model, we demonstrate that they do contain a massive dark component amounting to about twice the visible matter. Staying within the standard cosmological framework, this result most likely indicates the presence of large amounts of unseen, presumably cold, molecular [hydrogen] gas. This additional mass should be present in the disks of their progenitor spiral galaxies, accounting for a substantial part of the so-called missing baryons." (Science 25 May 2007 Vol.316 no.5828, pp.1166-1169)

Jerome Drexler originated the five-year-old relativistic-proton dark matter theory, model, and cosmology and disclosed and defended it in the form of a 32-slide Powerpoint presentation to two professors at the University of California, Santa Cruz campus in April 2003. He then expanded his presentation to 108 slides and transformed it into a 156-page paperback book, *How Dark Matter Created Dark Energy and the Sun,* which was published December 15, 2003.

PREREQUISITE CHAPTER F

Eroding High-Energy Dark Matter Particles in Galaxy Clusters May Explain the Universe's Acceleration

July 19, 2007 — The 2006 Shaw Prize in Astronomy and 2007 Gruber Cosmology Prize were awarded for the amazing discovery of the accelerating expansion of the universe, but the cause of this phenomenon is still considered a mystery.

Dark matter comprises about 83 percent of the mass of the universe. If dark matter were comprised of uncharged particles, the author would have no explanation for the accelerating expansion of the universe.

Ordinary matter of the universe is comprised of over 99 percent in the form of hydrogen and helium in the well-known atomic ratio of 12:1. The author, over the past six years, has accumulated overwhelming evidence that the dark matter of the universe is comprised of relativistic protons and helium nuclei in the nuclei ratio about 10:1 to 12:1 that orbit galaxies and groups of galaxies. The proton orbits are not determined by gravitational attraction as in the case of our solar system, but by the extragalactic magnetic field and the electric charges and the velocities of the protons and helium nuclei. More exactly, the proton orbits are determined by the

Larmor radius equation, rather than Kepler's laws, which apply to the solar system. (See Appendix II, Slides 85-87)

Based upon the overwhelming evidence provided by the author's astro-cosmology books of 2003, 2006, and 2008 and his scientific papers of 2005 and 2007, the reader is asked to proceed, at the moment, on the basis that the author's dark matter theory is probably valid. The following is the only plausible explanation, publicly proposed to date, for the mystery of the accelerating expansion of the universe.

Observed galaxy clusters are rushing apart faster and faster thereby accelerating the expansion of the universe. If the galaxies and the stars were similarly rushing apart faster and faster, then the *hypothetical* concept of *dark energy* pervading all space might be valid. But this isn't the case.

The galaxy clusters are indeed rushing apart as their separation velocities are accelerating, but the separation velocities of the galaxies and of the stars are not known to be accelerating. If the hypothetical pervasive *dark energy* actually existed, probably all three of these celestial bodies would be rushing apart faster and faster.

In an expanding universe galaxy clusters are moving away from one another according to Hubble's law. When the galaxy-cluster separation velocities continue to rise, as in our universe, the universe is in an accelerating expansion mode. Since 1998, this has been attributed to *dark energy*, called "the most profound mystery in all of science" by University of Chicago cosmologist Michael Turner.

In that year, two astronomical research studies led by Saul Perlmutter, of the Lawrence Berkeley National Laboratory and by Brian Schmidt of the Australian National University, of astronomical events involving exploding stars (supernovae) led to the award-winning discovery that the expansion of the universe was accelerating. The cause was attributed to *dark energy*, a hypothetical form of energy assumed to permeate *all* space and to have negative pressure resulting in a repulsive gravitational force.

In 2001, Michael Turner essentially removed the word *all* from this definition when he wrote, "Dark energy by its very nature is diffuse and a low-energy phenomenon. It probably cannot be produced at accelerators; it isn't found in galaxies or even clusters of galaxies"[78]. Astro-cosmology author Jerome Drexler has based his accelerating universe theory on Michael Turner's statement, "it isn't found in galaxies or even clusters of galaxies." That is, Drexler's criteria for the mysterious repulsive force are: it pushes galaxy clusters apart faster and faster, but it doesn't push galaxies or stars apart faster and faster.

Drexler's theory explains that, in a first phase, galaxy clusters, filled with relativistic dark matter protons orbiting groups of galaxies, are separating from each other with velocities proportional to their separations according to Hubble's law. Then, when galaxy clusters' relativistic dark matter protons erode relativistic mass at a sufficiently high rate, via synchrotron emission of extreme ultraviolet (EUV) or soft X-ray photons, the separation velocities of galaxy clusters will increase because of the reduction in the gravitational attraction between them and owing to the Law of Conservation of Linear Momentum. (The "sufficiently high rate" should be determined by the gravitational attraction between galaxy clusters versus the level of synchrotron emission from the dark matter protons.)

This increases cluster-to-cluster separation and further lowers the gravitational attraction between galaxy clusters, thereby further accelerating both galaxy cluster separation and the expansion of the universe. The force causing this accelerating separation of galaxy clusters is essentially only between galaxy clusters, not between galaxies, or between stars, as will be explained.

Drexler's dark matter candidate is comprised of relativistic protons and relativistic helium nuclei in a ratio of between 10:1 and 12:1. This is compatible with a recent scientific paper "Missing Mass in Collisional Debris from Galaxies" in the May 25, 2007 issue of Science magazine[27], which concludes, "The most natural [dark matter] candidate is molecular hydrogen in some hard-to-trace form."

The extragalactic magnetic fields cause the charged ultra-high-energy dark matter protons to remain within galaxy clusters and to radiate synchrotron emission primarily in the form of extreme ultraviolet (EUV), UV, or soft X-ray, or infrared photons.

Synchrotron emission from a proton is in the form of photons in its direction of motion. Since protons in a galaxy cluster are orbiting groups of galaxies, such emission from a galaxy cluster should be relatively isotropic with respect to the cluster's linear motion. The higher energy relativistic protons, orbiting groups of galaxies, would be emitting higher power synchrotron emission, causing them to lose energy and relativistic mass faster, as if their mass were eroding faster.

How can synchrotron emission push galaxy clusters apart without pushing galaxies apart? First of all, the power of synchrotron emission from a relativistic proton moving across a transverse magnetic field is directly proportional to the square of the proton's energy. Secondly, the energies estimated for the dark matter protons orbiting groups of galaxies in a galaxy cluster would be about 30 times higher than the energies calculated for the dark matter halo protons orbiting a single spiral galaxy, such as the Milky Way. Thus, the synchrotron emission power per proton and relativistic mass loss rate per proton are about 900 times greater for accelerating the separation velocities of galaxy clusters than for accelerating the separation velocities of galaxies.

Another question is why the accelerating expansion did not begin until about five billion years ago, as reported by astronomers. Perhaps in earlier epochs the much smaller separations of galaxy clusters and the inverse-square law of gravity led to a very much higher gravitational attraction between clusters that minimized the synchrotron-emission repulsive effect. Note that in the past 13.4 billion years galaxy-cluster separation distances have grown by a factor of the order of 1000.

Drexler's theories for dark matter and the accelerating cosmos could be tested by NASA in 2009-2010 (and Russia in 2010[46]) after the Hubble telescope's EUV/UV sensitivity is increased by a factor of 10. The detection of EUV or UV photons or soft X-rays from dark matter of our Local Group galaxy cluster would support Drexler's relativistic-proton dark matter theory.

Calculations indicate that synchrotron emission from the Milky Way's dark matter halo should have a broad peak in the infrared, while synchrotron emission from dark matter protons orbiting groups of galaxies in galaxy clusters should have a broad peak in the EUV or soft X-ray region.

Is the dark matter of the universe comprised of ultra-high-energy relativistic protons? Consider these five paragraphs.

The only plausible explanation for the *accelerating expansion of the universe*, announced to date, requires that the dark matter of the universe be comprised of ultra-high-energy relativistic protons *(see Chapter F)*.

The only plausible explanation for the *Big Bang satisfying the Second Law of Thermodynamics*, announced to date, requires that dark matter be comprised of ultra-high-energy relativistic protons *(see Chapter C)*.

The only announced plausible explanation for *UHE cosmic ray protons* with energies above 60 EeV bombarding Earth's atmosphere requires that dark matter be comprised of ultra-high-energy relativistic protons *(see Chapters G and H)*.

The only announced plausible explanation for *cosmic inflation* requires that dark matter be comprised of ultra-high-energy relativistic protons *(see Chapter H)*.

The only announced plausible explanation for the *cosmic web* requires that dark matter be comprised of ultra-high-energy relativistic protons *(see Chapter J)*.

PREREQUISITE CHAPTER G

Auger Collaboration Probably Detected Big-Bang-Created UHECR Protons After Their Ejection by Merging Galaxy Clusters

Nov. 26, 2007 — The Pierre Auger collaboration, an international project involving 370 scientists and engineers from 17 countries, announced on Nov. 8 the significant discovery of 27 ultra-high-energy cosmic ray (UHECR) protons emitted from some unknown extragalactic sources located within 3 degrees of "active galactic nuclei" (AGNs) within about 250 million light-years of Earth [72].

There was no explanation of the sources and enormous energies of these UHECR relativistic protons or of their acceleration means.

The following theory may explain the discovery: The Second Law of Thermodynamics required that the big bang, while creating the universe's mass and energy, generate primarily relativistic protons in order to minimize the entropy of the universe at the beginning of time. From the generally accepted big bang temperatures, some proton energies might have been at energy levels between 1 million EeV to 10 million EeV. Over the subsequent 13.7 billion years the so-called GZK proton-energy-loss effect (explained later) could diminish these proton energies by a factor of 100,000, yet would still permit the arrival at Earth of a small percentage of 10 EeV (10 exa-electronvolt) to 100 EeV ultra-high-energy cosmic ray protons, which could be observed today.

The author presents data and arguments that the recent Auger collaboration discovery of 60 EeV cosmic ray protons from sources "that lie within roughly 250 million light-years of Earth" probably represents these energy diminished big bang relativistic protons that were ejected from their long-term steady state orbital paths, around several to tens of spiral galaxies, by transient magnetic field shocks brought about by the merging of galaxy clusters.

The Nov. 8 University of Chicago news release contained the following four relevant quotations regarding the discovery of the Auger collaboration:

> "The international Auger collaboration has traced the rain of high-energy cosmic rays [UHECRs] that continually pelts the Earth to the cores of nearby galaxies, which emit prodigious quantities of energy."

> "In the next few years, our data will permit us to identify the exact sources of these cosmic rays and how they accelerate these particles."

> "Cosmic rays — mostly protons — fly through the universe at nearly the speed of light. The most powerful cosmic rays contain more than one hundred million times more energy than the particles produced in the world's most powerful particle accelerator."

> "Scientists have long considered Active Galactic Nuclei (AGN) to be possible sources of high-energy cosmic rays. And while

they have now found a strong correlation between the two, exactly what accelerates cosmic rays to such extreme energies remains unknown."

On Nov. 9, 2007, Science magazine published a scientific paper entitled, "Correlation of the Highest-Energy Cosmic Rays with Nearby Extragalactic Objects"[72].

The Nov. 9 Science magazine also published an article by Adrian Cho that contained the following four substantive quotations:

"The cosmic rays do not point precisely to the AGNs; presumably, our galaxy's magnetic field deflects them in transit. Details of the analysis suggest that the cosmic rays are protons."

"Physicists measure the energy of the highest energy rays in exa-electron volts EeV. The Auger team finds that rays with energies higher than 57 EeV — of which they see 27 — generally come from directions within 3° of "active galactic nuclei" (AGNs) that lie within roughly 250 million light-years of Earth."

"Meanwhile, theorists have a puzzle to solve: Exactly how might an AGN accelerate a proton to such mind-boggling energies?"

Reflecting Alan Watson's key comments, Adrian Cho wrote, "The results don't prove AGNs are sources of the rays. Anything else that's distributed on the sky in the same way as AGNs could be the source," Alan Watson says. He added, "For example, galaxies tend to clump, so some other sort of galaxy might be the culprit." (Alan Watson is a professor at the University of Leeds and co-founder of the Pierre Auger Observatory.)

A Nov. 9 ScienceDaily article added, "Galaxies that have an AGN seem to be those that suffered a collision with another galaxy or some other massive disruption in the last few hundred million years."

Google News Nov. 9: Comment by Dr. Paul M. Mantsch, Senior physicist at Fermilab and project manager of the Pierre Auger Project. A two-paragraph article included the key sentence, "Although violent AGN are good candidates for sources, they might only be tracers for some other kind of sources nearby."

The mysteries of the Auger collaboration discovery reminded Silicon Valley's inventor/scientist, Jerome Drexler, that he had posited a solution to a similar mystery in the summer of 2005 and later wrote about it in Chapter 2006-47 of his May 2006 book *Comprehending and Decoding the Cosmos.* Excerpts from that chapter, as follows, attempt to explain the Auger collaboration's significant discovery of UHECRs from the directions of AGNs without evoking black holes or AGNs:

> Then, on July 29, 2005, Elena Pierpaoli and Glennys Farrar posted a paper on the Physics arXiv, astro-ph/0507679 entitled, 'Massive galaxy clusters and the origin of Ultra High Energy Cosmic Rays', in which the massive galaxy clusters are described as a merging pair of clusters. In their paper, .Pierpaoli and Farrar suggest a possible explanation for the observed phenomenon [73] as follows:

A merging pair of clusters would be expected to have very large scale, strong magnetic shocks which could be responsible for accelerating UHECR even if there is no AGN (active galactic nuclei) or GRB (gamma ray burst) associated with the galaxy clusters.

Note that Pierpaoli and Farrar believe that lower-energy cosmic ray protons are accelerated into UHECRs through magnetic shocks created in the merging galaxy clusters.

Perhaps both research groups are correct in concluding that the UHECRs may have been accelerated. However, there is another possibility. During the pre-merger period, UHECRs, defined as having energies at or above 1 EeV (1 exa-electron volt), might have been orbiting galaxy clusters within their dark matter halos in a steady-state manner according to the Larmor Radius equation. Given the general size of galaxy clusters and the generally accepted magnitude range of the extragalactic magnetic field, one would conclude that most of the pre-merger orbiting dark matter protons at the periphery of galaxy clusters would be UHECRs. [See Appendix II, Slides #85-#87]

The galaxy cluster merging process would upset the steady-state Larmor orbiting symmetry of the UHECRs. The combining of the magnetic fields of the two merging spiral galaxy clusters could create transient magnetic field distortions (shocks), which would cause a number of UHECRs to be deflected off into space, with some being Earthbound. This theory might be called the deflection-from-orbit theory of UHECR emission. It is presented as a plausible alternative theory to the shock acceler-ation UHECR theory, which remains unproven according to the two research groups.

The Pierpaoli-Farrar paper indicates that the authors have found data about several UHECR events with energies at about 50 EeV departing from a merging pair of galaxy clusters observed in the SSDS DR3. In the 22 Nov 2005 (v3) abstract of the Pierpaoli-Farrar paper they say, "For cosmic rays with energies above 50 EeV the observed correlation is the strongest for angles of 1.2-1.6 degrees where it has a chance probability of about 0.1 percent."

Also in the v3 version of the paper, the Discussion and Conclusions section says, "Therefore we conclude that the correlations between AGASA UHECR and galaxy clusters do not seem to be driven by the presence of BL Lac or AGN within the galaxy clusters."

Presented here in the previous four paragraphs are (1) Drexler's posited deflection-from-orbit theory of UHECR emission (2) the theory's brief description based upon the Relativistic-Proton Dark Matter hypothesis, and (3) its experimental basis derived from the Pierpaoli-Farrar astronomical data and their conclusions. The task left to complete is the positing of how the relativistic protons orbiting groups of galaxies within a galaxy cluster obtained their energy levels above 50 EeV. The following logical steps should lead toward that result.

Let us now compare the probability of validity of Drexler's relativistic-proton dark matter hypothesis to the probability of validity of the hypothesis that an AGN mechanism can accelerate protons to energies 100 million times higher than the most powerful particle accelerator on Earth. Before we proceed with that task, let us think about the Pierpaoli-Farrar researchers who in 2005 discovered UHECRs emanating from merging galaxy clusters which do no not exhibit AGNs.

If it required a black hole-AGN accelerator mechanism to create the UHECRs for the Auger collaboration discovery, what other accelerator mechanism was used by the merging galaxy clusters reported by Pierpaoli-Farrar which did not possess AGNs? It is highly unlikely that two completely different accelerator mechanisms exist in the universe that could achieve the enormous proton accelerations described in the previous paragraph. In fact it is highly unlikely that any cosmological accelerator mechanism could achieve those proton accelerations except the big bang.

Drexler's scientific books and papers explain that the big bang created relativistic protons having energies up to 10 million times higher than 1 EeV UHECRs of today in order for the big bang to satisfy the Second Law of Thermodynamics. See Drexler's online published paper, physics/0702132, (or Chapter C) and his Dec.15, 2003 book, which state that the big bang generated protons at 10 million EeV. Then over the following 13.7 billion years their energies could decline by the so-called GZK loss mechanism (and synchrotron emission) by a factor of 100,000 and still be at the 100 EeV UHECR energy level observed on Earth on rare occasions.

The 1966 GZK cut-off theory has been used to argue that relativistic-proton dark matter could not. have survived 13.7 billion years of travel. The GZK cut-off theory predicts that because of the interaction with the cosmic microwave background, relativistic protons cannot have energies higher than 6×10^{19} eV at Earth since above those levels they would have lost energy rapidly in collisions with the CMB.

Some cosmologists, who have been inclined to use the 1966 GZK cutoff theory to rule out the relativistic proton dark matter model, were apparently not aware of the 1998 paper designated hep-ph/9808446 and entitled "Evading the GZK Cosmic-ray Cutoff"[75] by Sidney Coleman and Nobelist Sheldon L. Glashow or the 1997 published chapter of Nobelist James W. Cronin in *Unsolved Problems in Astrophysics*[80]. James Cronin's chapter in the book reported two observed cosmic ray protons with energies about four times higher than the theoretical GZK proton-energy cutoff. The presence of UHE cosmic ray protons in the solar system, including a very few with energies well above the GZK cutoff, adds some plausibility to an anti-GZK effect.

Also, researchers have reported an anti-GZK effect that arises when UHE relativistic protons moving through an intergalactic magnetic field experience diffusive propaga-tion. They report that this effect causes a jump-like increase in the distance UHE protons can travel. From its description in the literature, apparently relativistic protons orbiting a galaxy in its halo's magnetic field also would experience an anti-GZK effect. For example, see astro-ph/0507325 authored by R. Aloisio and V. S. Berezinsky [79].

It should be noted that there is a possibility that the GZK proton energy loss effect could be much lower for protons orbiting galaxies in a galaxy cluster[79] than those protons passing through extragalactic space. Physics Nobel Laureate Sheldon L. Glashow, et al, of Harvard wrote a paper [75] on a related subject in 1998 entitled, "Evading the GZK Cosmic-Ray Cutoff." The abstract reads as follows.

"Explanations of the origin of ultra-high energy cosmic rays are severely constrained by the Greisen-Zatsepin-Kuz'min [GZK] effect, which limits their propagation over cosmological distances. We argue that possible departures from strict Lorentz invariance, too small to have been detected otherwise, can affect elementary-particle kinematics so as to suppress or forbid inelastic collisions of cosmic-ray nucleons with background photons. Thereby can the GZK cutoff be relaxed or removed."

What is Lorentz invariance? This paragraph, from an October 2005 Discover magazine [76] helps explain it. "If two physicists traveling freely through empty space passed by each other at a high relative velocity, we couldn't tell in any universal sense which one was stationary and which was moving — it's all relative, if you like. If we violated Lorentz invariance by having a vector field get a nonzero value in the vacuum, we *could* tell who was stationary and who was moving — the vector would define a preferred rest frame."

To a relativistic proton, a magnetic field region should be a vector field that probably departs from Lorentz invariance since the proton's orthogonal acceleration would be determined by its direction of motion relative to the magnetic field lines. (The *diffusive propagation* mentioned above also implies a vector field.) Furthermore, the galactic magnetic field strength, such as that of the Milky Way, about 2000 times stronger than the magnetic field of extragalactic space, may substantially increase the departure from Lorentz invariance. Thus we should expect that relativistic protons orbiting galaxies might evade the GZK cutoff while protons racing through extragalactic space probably would be subjected to it.

Also, the terms "GZK cutoff" and "GZK limit" are misleading. The terms imply 60 EeV UHECRs are capable of traveling only about 300 million light years through space, but in actuality that *cutoff* represents only a very large percentage energy decline for the very highest energy UHECRs and a much smaller decline for the lowest energy UHECRs. Is there an actual cosmic-ray distance cutoff?

Further, it should be noted that under the relativistic proton dark matter theory/cosmology, the only relativistic protons that could have energies close to the theoretical GZK cutoff are those orbiting a galaxy supercluster. For this case, the proton flux density would be extremely low *(see Appendix II, Slide #17)*. For example, cosmic ray protons at an energy level of 10^{19} eV or above striking Earth's atmosphere total only 3 to 4 per square kilometer per century.

When the reader becomes convinced that the relativistic-proton dark matter is a tangible concept and the big bang satisfied the Second Law of Thermodynamics through a *violent radial dispersion of relativistic baryons*, then the deflection-from-orbit theory of UHECR emission, posited by Drexler above can be applied to the Auger discovery.

Note that a corollary emanates from the above presentation: The Auger collaboration discovery itself may represent new and important evidence supporting the validity of relativistic-proton dark matter and/or the relativistic big bang.

PREREQUISITE CHAPTER H

Nobel Laureates' Queries Point Toward Drexler's Dark Matter Theory and *Postmodern Cosmology*

December 11, 2007 – "How inflation happened a split second after the big bang." "Identify the exact sources of these cosmic rays and how they accelerate particles." Two Nobel Laureates in physics publicly raised these two unrelated queries in November and December 2007.

There is the possibility that plausible answers or helpful responses to both of these queries can be derived from a recent cosmology scientific paper utilizing Drexler's dark matter theory and big bang cosmology. The paper was published online on the physics arXiv on Feb. 15, 2007 as e-print No. physics/0702132. It is titled "A Relativistic-Proton Dark Matter Would Be Evidence the Big Bang Probably Satisfied the Second Law of Thermodynamics".

The paper argues that the big bang, which occurred at the beginning of time, must have satisfied the Second Law of Thermodynamics. Thus, immediately after the big bang the entropy of the universe would be at the lowest level it would reach throughout all time. Therefore the big bang should not be characterized, as it has been for over 40 years, as a chaotic fireball explosion associated with a high level of disorder and high entropy.

The very low entropy could be achieved by the big bang firing out, in all directions, high-velocity ultra-high-energy (UHE) relativistic protons and helium nuclei close to the well-known nuclei ratio of 12:1. In other words, the big bang could be characterized as a *violent radial dispersion of relativistic baryons.*

A very high percentage of their energies would be available to do work since their entropy, a measure of the percentage of their energy unavailable to do work, would be very low. Such a big bang, characterized by a *violent radial dispersion* of UHE relativistic nuclei, would be highly efficient and could create very high usable energy and very low entropy, and might be designated a *Relativistic Big Bang.* This concept is fundamental to Drexler's dark matter theory and his *Postmodern Cosmology.*

The *Relativistic Big Bang* would have the protons and helium nuclei being fired out at near the speed of light in almost a purely radial outward direction for a short-time first phase, followed by a second phase during which the magnetic field deflections and electric-charge repulsion of the particles would impart a transverse motion and angular momentum to the particles, thereby greatly reducing their radial outward velocities.

The *Cosmic Inflation* period could be related to the extremely short-time first phase of almost a purely radial outward motion near the speed of light of the relativistic protons and helium nuclei, beginning a split-second after the big bang created them.

Hopefully, this explanation will be considered a plausible answer or helpful response to a Nobel Laureate's December 2007 public query, "How inflation happened a split second after the Big Bang."

Let us consider the query about cosmic rays "identify the exact sources of these cosmic rays and how they accelerate these particles." The astronomical data used to arrive at answers to this two-part cosmic-ray query will be taken from the reports of the Pierre Auger collaboration, an international project involving 370 scientists and engineers from 17 countries[72] (*see also Chapter G*). They announced on Nov. 8, 2007 the significant discovery of 27 Ultra-High-Energy Cosmic Ray (UHECR) protons with energies higher than 57EeV. These are extremely rare events requiring a relativistic-proton detection system the size of Rhode Island.

The Second Law of Thermodynamics required that the big bang, while creating the universe's mass and energy, generate most of the mass in the form of relativistic protons and helium nuclei in order to minimize the entropy of the universe at the beginning of time.

From the generally accepted big bang temperatures, some proton energies might have been at energy levels between 1 million EeV to 10 million EeV. Over the subsequent 13.7 billion years the so-called GZK proton-energy-loss effect could diminish these proton energies by a factor of 100,000, yet would still permit the arrival at Earth of a small percentage of 10 EeV to 100 EeV UHE cosmic ray protons.

The author believes that the recent Auger collaboration discovery of 27 higher-than 57 EeV cosmic ray protons probably represents the energy-diminished big bang relativistic protons that over billions of years as stragglers in space finally found a home orbiting several to tens of galaxies. Later, they were deflected and ejected from their long-term steady-state orbital paths, around several to tens of spiral galaxies, by transient magnetic field shocks brought about by the merging of two galaxy clusters.

Also, the terms GZK cutoff and GZK limit are misleading. The terms imply 60 EeV UHECRs are capable of traveling only about 300 million light years through space, but in actuality that *limit* probably represents only a very large percentage decline for the very highest energy UHECRs and a very much smaller percentage decline for the lowest energy UHECRs. See Chapter G about cosmic ray protons, acceleration means, and anti-GZK effect.

PREREQUISITE CHAPTER J

In the Cosmic Web, Long, Large Filaments of Dark Matter (Announced By NASA 9/9/04) Form Galaxy Clusters Where They Intersect/Collide

A September 9, 2004 news release from NASA (and Harvard) entitled, "Motions in nearby galaxy cluster reveal presence of hidden superstructure" [33], regarding Chandra x-ray images of the Fornax cluster, states:

> Astronomers think that most of the matter in the universe is concentrated in long large filaments of dark matter and that galaxy clusters are formed where these filaments intersect.

The researchers' related paper astro-ph/0406216 is entitled, "The Chandra Fornax Survey - I: The Cluster Environment" [77]. This astronomically established filamentary description of dark matter in the Cosmic Web [43] appears to be much more compatible with the relativistic proton dark matter theory than the cold dark matter theory. It seems highly unlikely that the dark matter filamentary structure could be created by very slow moving, weakly interacting (only through gravitational tidal forces) particles.

The vision of dark matter filaments crisscrossing the cosmos gives the impression of high-velocity particles, while the crashing of intersecting dark matter filaments creating galaxy clusters gives the impression of a top-down theory of galaxy formation. Both of these impressions point toward and lend support to the relativistic dark matter theory/cosmology.

Furthermore, the theoretical WIMPs, being non-baryonic, cannot be transformed into hydrogen and helium where the dark matter filaments intersect/collide, whereas the relativistic protons and helium nuclei, being baryonic, can provide and feed hydrogen and helium to the galaxy clusters where the filaments intersect/collide.

For the above reasons, the September 2004 reports of the intersecting dark matter filaments seemed to be very supportive of Drexler's relativistic-proton dark matter theory and encouraged him to write his May 2006 book [35], *Comprehending and Decoding the Cosmos.*

APPENDIX II

Presented here are 18 selected pages involving six important references from J. Drexler's book *How Dark Matter Created Dark Energy And The Sun* (Universal Publishers Boca Raton, Florida USA, 2003).

Energies of Relativistic Protons Versus Their Relativistic Mass – Where Do They Exist in Nature?

Energy 9.38×10^8 eV	Relativistic Mass of a Proton In Terms of Its Rest Mass, m_o m_o
10^{10} eV	11 m_o
10^{11} eV	110 m_o
10^{12} eV	1,100 m_o
10^{13} eV	11,000 m_o
10^{14} eV	110,000 m_o
10^{15} eV	1,100,000 m_o
10^{16} eV	11,000,000 m_o
10^{17} eV	110,000,000 m_o
10^{18} eV	1,100,000,000 m_o

$$\text{Relativistic Mass} = \frac{\text{Energy (in joules)}}{C^2 \text{ (in meters/sec)}}$$

Page 22 of *How Dark Matter Created Dark Energy And The Sun*

SLIDE #15

Such Highly Energetic Protons Can Be Found Striking the Earth's Atmosphere as Cosmic Ray Protons

Approximate Kinetic Energy	Approximate Cosmic Ray Flux On the Earth's Atmosphere
10^8 to 10^{10} eV	Slightly less than 1,000 particles per square meter per second
10^{11} eV	One particle per square meter per second
7×10^{15} eV	One particle per square meter per year
3×10^{18} eV	One particle per square kilometer per year
10^{19} eV	3 to 4 particles per square kilometer per century

Page 23 of *How Dark Matter Created Dark Energy And The Sun*

SLIDE #16

Cosmic-Ray Energy Distribution
at the Earth*

CERN Courier, **Vol. 35, No. 10, December 1999**

[See Slide #16 for the Key Data Points]

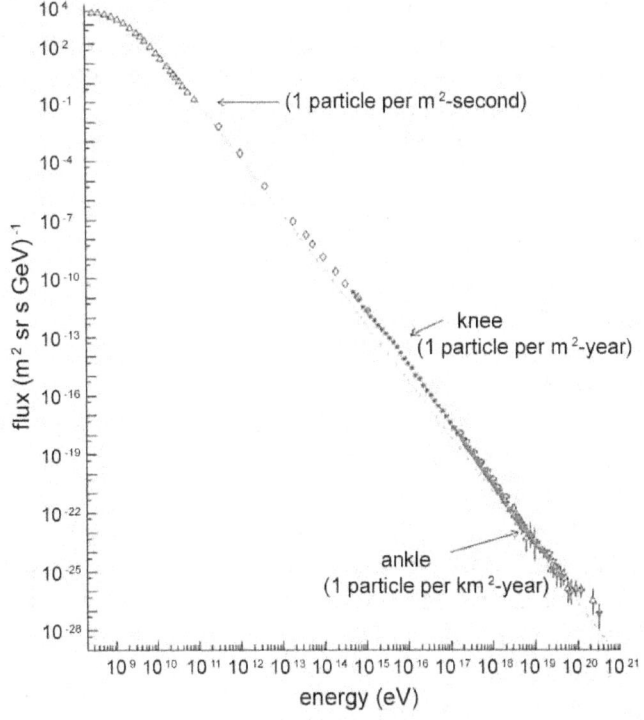

"The cosmic-ray energy distribution shows remarkable uniformity over 10 orders of magnitude. However, there are two kinds. The ACCESS experiment is designed to investigate 'the knee' (near 10^{15} eV)."

*Available on the Internet at http://www.cerncourier.com/main/article/39/10/8/1.

Page 24 of *How Dark Matter Created Dark Energy And The Sun*

SLIDE#17

The Accelerating Expansion Between Galaxy Clusters (Not Between Galaxies)

- Note that Michael Turner points out (see slide #54) that the dark energy "isn't found in galaxies or even clusters of galaxies." This is an extremely important point. This statement means that dark energy pushes galaxy clusters apart but doesn't push galaxies apart within clusters and doesn't push stars apart within a galaxy.

- This is observed locally. While the Universe is expanding, in our Local Group of galaxies Milky Way and Andromeda are moving towards each other at 119 km/sec. Also, our Local Group is moving toward the Local Supercluster at 600km/sec. This illustrates that local strong gravitational forces can overcome the dark energy antigravity forces. This also suggests that perhaps in the earlier, smaller, and denser Universe, when the Universe was less than about 8.2 billion years old, the closeness of the galaxy clusters favored dark matter gravity over dark energy antigravity, and expansion acceleration could not take hold. When the Universe was 8.7 billion years old, the expansion acceleration began, according to Adam Riess. (See slide #53.)

Page 65 of *How Dark Matter Created Dark Energy And The Sun*

SLIDE #58

Jerome Drexler's Theory of the Accelerating Expansion Between Galaxy Clusters

- As the synchrotron radiation emission of gamma rays continues, not only does the kinetic energy of the halo UHE protons fall, but their relativistic mass will fall as well (slides #39 and #40). See slide #15 and assume that the average kinetic energy of the UHE protons declined from 10^{16} eV to 5×10^{15} eV over a period of time in the dark matter halo of some galaxy cluster. This represents a decline in the dark matter mass in the halo of that galaxy cluster of 50% and perhaps a 40% decline in the mass of the combined galaxy cluster and its halo.

- Such a reduction in the dark matter halo mass around galaxy dusters would:

  (1) Raise the galaxy clusters' velocities under the Law of Conservation of Linear Momentum. (See footnote on slide #75.) [Drexler]

  and

  (2) Reduce each galaxy cluster's gravitational attraction to nearby galaxy clusters, thereby facilitating their more rapid separation. [Drexler]

Page 66 of *How Dark Matter Created Dark Energy And The Sun*

SLIDE #59

Drexler's Theory of the Accelerating Expansion Between Galaxy Clusters

- The gravitational attraction effect of item (2) on the previous slide will diminish through the years as the nearby galaxy clusters become more distant. Note from slide #54 that from the age of 300,000 years until today, the spacings between galaxies increased by a factor of 1,000.

- In an expanding Universe, all galaxy clusters are moving away from each other. Meanwhile, the masses of their dark matter halos of UHE protons are declining because of the synchrotron radiation energy losses. As a result, the velocity of every galaxy cluster should rise (owing to the reduced gravitational attraction between them and the Law of Conservation of Linear Momentum), thereby accelerating the expansion of the Universe. [Drexler]

Page 67 of *How Dark Matter Created Dark Energy And The Sun*

SLIDE #60

Drexler's Theory of the Accelerating Expansion Between Galaxy Clusters

- The antigravity repulsion between galaxy clusters is proportional to the relative decrease in mass of their dark matter halos of UHE protons owing to the protons' loss of kinetic energy through synchrotron radiation. [Drexler]

- Gravitational attraction is directly proportional to the product of the masses of two galaxy clusters and inversely proportional to the square of the distance between them.

- Thus, for the earlier, smaller, and denser Universe more than five billion years ago, the smaller distances between galaxy clusters caused the gravitational attraction between them to be high because of the inverse square of those smaller distances between galaxy clusters. This inverse-square relationship may be a principal reason that the accelerated expansion did not begin until five billion years ago when the conservation-of-momentum effect and reduced galaxy cluster mass finally overcame the gravitational attraction between galaxy clusters. [Drexler]

Page 68 of *How Dark Matter Created Dark Energy And The Sun*

SLIDE #61

Drexler's Theory:
UHE Cosmic-Ray Nuclei May Have Facilitated the Triggering of the Sun's Fusion Reaction

- With the UHE cosmic ray protons having a kinetic energy in the range of 10^{10} to 10^{20} electron volts, they not only added mass and fuel to the formation of the Sun but considerable nucleus-to-nucleus collision energy as well.

- These UHE nuclei provide a clue that the UHE protons and heavier high-collision-cross-section UHE nuclei may have facilitated the triggering of the Sun's fusion reactions and its birth.

- The traditional theory of the Sun formation involving a hydrogen gas cloud, forces of gravity, compression, and high temperature heating may have to be modified. (For further information about the UHE cosmic-ray nuclei, see slide #17, an energy distribution graph that is a plot of cosmic-ray particle flux versus particle energy.) [Drexler]

Page 78 of *How Dark Matter Created Dark Energy And The Sun*

SLIDE #71

What is the Difference Between a UHE Proton and a Cosmic-Ray Proton Bombarding the Earth?

- When a UHE proton in the halo of the Milky Way (a spiral galaxy) loses a significant portion of its kinetic energy over billions of years through synchrotron radiation, the proton will eventually plummet into the galaxy, thereby accelerating its energy loss. It then becomes one of the cosmic ray protons which bombard wide regions of the galaxy, including the solar system, and may have played a role in creating the Sun and other stars as explained in slides #62 – #71.

- With UHE protons remaining active for billions of years, they may be thought of as "immortal" UHE protons while at end of their life they become "mortal" cosmic ray protons plummeting into the galaxy in what I call a "death spiral."

Page 85 of *How Dark Matter Created Dark Energy And The Sun*

SLIDE #78

The Transformation of "Immortal" UHE Protons into "Mortal" Cosmic-Ray Protons Through the "Death Spiral"

- The synchrotron radiation loss of a relativistic charged particle is inversely proportional to both the radius of curvature of its path and the fourth power of its mass.

- The radius of curvature of a UHE proton's spiral path is equal to the Larmor Radius (see slide #85) and is directly proportional to the kinetic energy of the proton and inversely proportional to the magnetic field strength.

- The extragalactic magnetic field is reported to be about 1×10^{-9} gauss while the magnetic field in the interior of the Milky Way is about 2,000 times greater, at 2×10^{-6} gauss.

- Therefore, in the extragalactic dark matter halo of the galaxy, the magnetic field is very weak, the kinetic energy of the protons is high, the synchrotron radiation losses are extremely low, and the UHE proton may be able to circulate for billions of years. [Drexler]

Page 86 of *How Dark Matter Created Dark Energy And The Sun*

SLIDE #79

The Transformation of "Immortal" UHE Protons into "Mortal" Cosmic-Ray Protons Through the "Death Spiral"

- After billions of years in the extragalactic halo of a spiral galaxy, some of the UHE protons should eventually lose enough energy that their spiral paths will be reduced in diameter and the UHE protons will approach the surface of the galaxy. When the UHE halo protons enter the galaxy, their energy has perhaps diminished by a factor of 10 or so and the magnetic field might be about 100 times greater, thereby increasing the synchrotron radiation loss by a factor of 1,000. (The radius of a proton's spiral path is called the proton's Larmor Radius, which can be calculated as shown in slide #85.)

- Thus, as a UHE proton enters the galaxy, its energy will plummet further, say to half, thereby doubling the synchrotron radiation loss. The proton's energy will drop more and more, and it will enter into the "death spiral" as it plunges deeper into the galaxy and begins to be known as a cosmic-ray proton.

Page 87 of *How Dark Matter Created Dark Energy And The Sun*

SLIDE #80

The Transformation of "Immortal" UHE Protons into "Mortal" Cosmic-Ray Protons Through the "Death Spiral"

- The rate of synchrotron radiation is inversely proportional to the fourth power of the mass of the particle. Therefore, synchrotron radiation from protons is infinitesimal compared to synchrotron radiation from electrons. More precisely, proton synchrotron radiation losses are lower than radiation losses from electrons, following the same radius of curvature path, by a factor of about 11 trillion (the proton/electron mass ratio of 1,836 to the fourth power).

- The discussion on slides #79 and #80 has to do with spiral galaxies, which are known to have a dark matter halo and also a black hole containing a few million solar masses.

Drexler's Cosmic-Ray Cosmology
Applied to Galaxy Formation

The Proton Larmor Radius

- A proton crossing an orthogonal magnetic field enters into a spiral path. The radius of one cycle of that spiral path is called the Larmor Radius.

- The Larmor Radius of a proton can be calculated as:

$$r = 110 \text{ Kpc} \times \frac{10^{-8} \text{ gauss}}{B} \times \frac{E}{10^{18} \text{ eV}}$$

  where Kpc means kilo parsec and
  one parsec equals 3.26 light-years

- The galactic magnetic field within the Milky Way is approximately 2×10^{-6} gauss compared to the extragalactic magnetic field at 1×10^{-9} gauss.

Page 92 of *How Dark Matter Created Dark Energy And The Sun*

SLIDE #85

Drexler's Cosmic-Ray Cosmology
Applied to Galaxy Formation

The Proton Larmor Radius

- The Larmor Radius for a 10^{16} eV proton in the Milky Way halo's extragalactic magnetic field of 10^{-9} gauss is 11 Kpc; for a 10^{17} eV proton it is 110 Kpc; and for a 10^{18} eV proton it is 1,100 Kpc.

- The diameter of the Milky Way galaxy is about 100,000 light-years or 30.7 Kpc and its radius is about 15 Kpc.

- Studies in 1999 found that the dark matter halo of a spiral galaxy extends about 10 to 20 times the size of the visible regions (slide #3). Using a factor of 15, the radius of the dark matter halo would extend to perhaps 225 Kpc.

- Thus, some 10^{16} eV protons with a Larmor Radius of 11 Kpc might stay within the Milky Way galaxy's 15 Kpc radius. A 10^{17} eV proton with a Larmor Radius of 110 Kpc might remain within the halo's 225 Kpc radius, but a 10^{18} eV proton would probably leave both the galaxy and the halo.

Page 93 of *How Dark Matter Created Dark Energy And The Sun*

SLIDE #86

Only One Dark Matter Candidate Establishes the Approximate Size of the Milky Way

- From slides #85 and #86 it has been shown that the Larmor Radius for a 10^{16} eV proton in the Milky Way's extragalactic magnetic field of 10^{-9} gauss is 11 Kpc; for a 10^{17} eV proton it is 110 Kpc; and for a 10^{18} eV proton it is 1,100 Kpc.

- In slide #17 entitled, "Cosmic-Ray Energy Distribution for the Milky Way," the "knee" of the curve falls on a proton energy of approximately 5×10^{15} eV, which means that some protons with that energy might stay within the galaxy.

- From the above it could be concluded that the Milky Way galaxy should have a minimum radius of about 11 Kpc, compared to astronomers' estimate of about 15 Kpc.

- This seems to provide additional evidence that UHE protons are a credible dark matter candidate. No other currently proposed dark matter candidate can be used to estimate the size or even the order of magnitude of the size of the Milky Way galaxy.

Page 94 of *How Dark Matter Created Dark Energy And The Sun*

SLIDE #87

Drexler's Cosmic-Ray Cosmology
Applied to Galaxy Formation

Some Plausible Speculations

- When a UHE proton moving along a certain path crosses orthogonal magnetic field lines, it is deflected up or down depending upon the magnetic field direction. The degree of deflection is proportional to the orthogonal magnetic field strength.

- Any magnetic deflection of a UHE proton reduces its velocity in the direction of its original path for two reasons: the deflection itself will cause its direction to change, and synchrotron radiation losses will reduce its forward velocity.

- Thus, if UHE protons that are moving through space at a certain velocity encounter a bulge (increase) in the magnetic field strength, they will not pass through the magnetic bulge region as quickly as through a no-bulge region and could possibly linger in the region. To describe this magnetic attraction effect for relativistic protons and electrons, I will use the term "attract" in quotes.

Page 95 of *How Dark Matter Created Dark Energy And The Sun*

SLIDE #88

Drexler's Cosmic-Ray Cosmology
Applied to Galaxy Formation

Some Plausible Speculations

- Since electrons lose energy through synchrotron radiation at a rate 11 trillion times faster than protons, electrons would lose their energy quickly and tend to circulate and accumulate in magnetic-bulge regions.

- Such electron-filled regions would add a coulomb attractive force to UHE protons, slowing them down and also facilitating their conversion into hydrogen atoms. Conceivably, this process could lead, in some cases, to negatively charged proto-galaxies or galaxies surrounded by positively-charged UHE proton dark matter. This, in turn, could lead to lightning-like proton electrical discharges creating a multitude of gamma-ray bursts (GRBs) of immense proportions.

- As the UHE protons slow down in close proximity to a magnetic-field bulge they could linger close by, adding mass and gravitational attraction to the magnetic bulge region.

Drexler's Cosmic-Ray Cosmology
Applied to Galaxy Formation

Some Plausible Speculations

- Magnetic-field bulges "attract" relativistic electrons and protons. The relativistic proton and electron "attraction" to a magnetic bulge may be different from, but as real as, the gravitational attraction between two masses.

- Previously, my concept of the "death spiral" was discussed with regard to UHE protons plummeting into the galaxy from the halo. In that case the galaxy magnetic field was about 2,000 times the strength of the extragalactic field. The Milky Way is an example of a significant magnetic-field bulge covering a large area in space that should capture almost all UHE protons spiraling through. In the solar system, the Sun is an example of a very significant magnetic-field bulge. With its magnetic field strength of about 50 gauss and the Milky Way's magnetic field strength at 2×10^{-6} gauss, the magnetic field ratio is 25 million.

REFERENCES

References for *Our Universe via Drexler Dark Matter*, the fourth book of Drexler's quadrology

1. G. Blumenthal, S. Faber, J. R. Primack, and M. J. Rees, 1984, *Nature* 311, 517.

2. L.Chuzhoy and E.W.Kolb, 2008, "Reopening the Window on Charged Dark Matter," arXiv:0809.0436 v1.

3. R. Courtland, "Is dark matter a wimp or a champ?", *New Scientist* 9 September 2008.

4. H.Johnston, "Galaxy survey casts doubt on cold dark matter", http://physicsworld.com/cws/article/news/36372 .

5. M. J. Disney, et al, "Galaxies appear simpler than expected", 2008 *Nature* 455, 1082-1084.

6. M. J. Disney, "Modern Cosmology:Science /Folk tale?" http://www.americanscientist.org/template/AssetDetail/a ssetid/55839?&print=yes .

7. J. Drexler, *Discovering Postmodern Cosmology* Ch. 24 (Universal Publishers, Boca Raton, Florida, 2008).

8. F. Bournaud, "Missing Mass in Collisional Debris from Galaxies", *Science* 25 May 2007, Vol.316 no.5828, p.1166-1169.

9. J. Drexler, *Discovering Postmodern Cosmology*, Ch. 19, (Universal Publishers, Boca Raton, Florida, 2008).

10. G.Gilmore, M.I.Wilkinson, et al 2007, "The Observed properties of Dark Matter on small spatial scales", arXiv:astro-ph/0703308v1.

11. J. Drexler, 2007, "A Relativistic-Proton Dark Matter Would Be Evidence the Big Bang Probably Satisfied the Second Law of Thermodynamics", physics/0702132.

12. G. R. Farrar and R. A. Rosen, 2006, "A New Force in the Dark Sector?", arXiv: astro-ph/0610298,

13. C. H. Gibson, 2006, "Cold Dark Matter Cosmology Conflicts with Fluid Mechanics", astro-ph/0606073.

14. Ria Novosti, 2006, "Russia to launch unique space observatory", according to Professor Boris Shustov, http://en.rian.ru/analysis/20060620/49784754-print.html .

15. J. Drexler, 2005, "Identifying Dark Matter through the Constraints Imposed by Fourteen Astronomically Based 'Cosmic Constituents'", astro-ph/0504512 v1.

16. A.de Rújula, S.L.Glashow, and U.Sarid, March 1990, "Charged dark matter", *Nuclear Physics*. B, Particle Physics, Volume 333, Issue1, p 173-194.

17. J.L. Feng and J. Kumar, "Dark-Matter Particles without Weak-Scale Masses or Weak Interactions," *Phys. Rev. Lett.* 101, 231301 (2008) [4 pages]

18. A.Kogut, et al "ARCADE 2 Observations of Galactic Radio Emission", http://arxiv.org/abs/0901.0562 v1

19. A. Dekel, et al, "Cold streams in early massive hot haloes as the main mode of galaxy formation", 2009 *Nature* 457, 451-454.

20. Cosmic-Ray Energy Distribution at the Earth, http://www.cerncourier.com/main/article/39/10/8/1. *CERN Courier*, Vol. 35, No. 10, December 1999.

21. ScienceDaily (Jan.8, 2009), "NASA Space Balloon Mission Tunes In To Cosmic Radio Mystery," http://www.sciencedaily.com/releases/2009/01/0901071725 46.htm .

22. D.A. Thilker,et al, "Massive star formation within the Leo 'primordial' ring, " 2009 Nature 457, 990-993.

23. R.Cowen, Science News (Feb. 18, 2009), "Galaxy Mix: No Dark Matter Required – Dark matter may not http://www.sciencenews.org/view/generic/id/41011/title/ Galaxy_mix_No_dark__matter_required

24. T. Stanev, *High Energy Cosmic Rays* (Springer-Verlag Berlin Heidelberg New York, 2004).

25. A. Vikhlinin, et al "Chandra Cluster Cosmology Project III: Cosmological Parameter Constraints," *The Astrophysical Journal*, Volume 692, Issue 2, pp. 1060-1074 (2009).

26. R. Cowen, Science News (Jan. 3, 2009) "Dark Energy Constantly with Us" - http://www.sciencenews.org/view/generic/id/39341/title/Da rk_energy_constantly_with_us .

27. P. Sokolsky, et al, "First Observation of the Greisen-Zatsepin-Kuzmin Suppression", *Phys. Rev. Lett.* 100, 101101 (2008) [5 pages].

28. C. A. Collins, et al, " Early assembly of the most massive galaxies", 2009 *Nature* 458, 603-606.

29. R. Cowen, Science News (April 25, 2009) "Heavyweight Galaxies In The Young Universe-Newfound massive galaxies",
http://www.sciencenews.org/view/generic/id/42419/title/ Heavyweight_galaxies_in_the_young_universe .

30. K.Glazebrook et al, "A high abundance of massive galaxies 3-6 billion years after the Big Bang", 2004, *Nature*, 430, 181.

31. A.Cimatti et al, "Old galaxies in the young universe", 2004, *Nature,* 430, 184.

32. P.J. McCarthy, D.L. Borgne, D.Crampton, H-W Chen, R.G. Abraham, K. Glazebrook et al. 2004, astro-ph/0408367 v1

33. "Motions in nearby galaxy cluster reveal presence of hidden superstructure", NASA Marshall Space Flight Center Release No.04-231, Huntsville, Alabama, September 9, 2004, p.1.

34. J. Drexler, *How Dark Matter Created Dark Energy and the Sun,* (Universal Publishers, Parkland, Florida, 2003).

35. J. Drexler, *Comprehending and Decoding the Cosmos,* (Universal Publishers, Boca Raton, Florida, USA, 2006).

36. J. Drexler, *Discovering Postmodern Cosmology,* (Universal Publishers, Boca Raton, Florida, 2008).

37. R. Cowen, Science News (May 4, 2009) "Introducing the Young Milky Ways — Astronomers discover ancestors…",
http://www.sciencenews.org/view/generic/id/43438/title/Int roducing_the_young_Milky_Ways

38. Carnegie Institution for Science (April 22, 2009) "Mysterious Space Blob Discovered at Cosmic Dawn", http://www.ciw.edu/news/mysterious_space_blob_discover ed_cosmic_dawn

39. AOL news (April 22, 2009) "Experts Puzzled by Strange Space Blob", http://news.aol.com/article/space-blob/443146 .

40. M. Ouchi, A. Dressler et al, "Discovery Of A Giant LYα Emitter Near..." arXiv:0807.4174v2 21 Feb. 2009, *The Astrophysical Journal*, Volume 696, pp. 1164-1175 (2009).

41. J.Achenbach, Washington Post, May 16, 2009, "Astronauts Successfully Complete Tricky Repair on Hubble", http://www.washingtonpost.com/wp-dyn/content/article/2009/05/16/AR2009051600330.html .

42. D.Perlman, Chronicle Science Editor, "Hubble probing mysteries of deep space," http://www.sfgate.com/cgi-bin/article.cgi?f=/c/a/2009/05/24/BARJ17ON67.DTL .

43. R. A. Ibata and G. F. Lewis, "The Cosmic Web in Our Own Backyard", Science, 4 January, 2008, Vol.319. No. 5859 p. 50-52.

44. "NASA Announces Details of Hubble Servicing Mission" Jan.8, 2008, http://www.reuters.com/article/pressRelease/idUS169689 +08-Jan-2008+PRN20080108 .

45. F. Zwicky, 1937, *Astrophys. J.* (Lett) 86, 217.

46. RIA Novosti, (20/06/2006) "Russia to launch unique [UV] space observatory [in 2010]", http://en.rian.ru/analysis/20060620/49784754.html .

47. Cheng-Jiun Ma, et al, 2009 "An X-ray/Optical Study of the Complex Dynamics of the Core of the Massive Intermediate-Redshift Cluster MACSJ0717.5+3745", *The Astrophysical Journal* (Lett) Volume **693**, L56-L60.

48. newscientist.com., 4 May 2009 "Dark matter 'highway' funnels gas into galactic pileup", http://www.newscientist.com/article/dn17074-dark-matter-highway-funnels-gas-into-galactic-pileup.html

49. Scientific American, 2009, "Majestic Universe", SCA50967

50. Queen's News Centre, June 9, 2009, CFI (Canadian Found. Innovation) awards Queen's researchers $42.7 M, http://qnc.queensu.ca/story_loader.php?id=4a3b7f9d4762a .

51. The Sudbury Star-Ontario, Canada, June 2009, "Government of Canada Invests in SNOLAB", http://www.thesudburystar.com/Community/NewsDisplay.aspx?c=19460 .

52. SpaceDaily.com, Paris, France, June 29, 2009, The "Invisible Universe" Exhibition/Symposium, http://www.spacedaily.com/reports/The_Invisible_Universe_Exhibition_Symposium_999.html ,

53. T. S. Kuhn, *The Structure of Scientific Revolutions*, Third Edition (The University of Chicago Press, 1996).

54. E. Masood, Science writer, guardian.co.uk Science Blog, "Are we witnessing the end of science?", http://www.guardian.co.uk/science/blog/2009/jun/22/end-science-unified-theory-mavericks .

55. M. Chown, newscientist.com, (8 July 2009) "Phantom menace to dark matter theory", http://www.newscientist.com/article/mg20327164.200-phantom-menace-to-dark-matter-theory.html .

56. M.Milgrom, 2009, "MOND effects in the inner solar system", arXiv.org >astro-ph> arXiv:0906.4817.

57. P. B. De Selding, Space.com, 23 June 2009, "NASA Poised to Join Europe's Mars Rover Mission", http://www.space.com/businesstechnology/090623-esa-mars-rover.html .

58. Innovations-Report, (14 July 2009) "Research sheds light on early star formation", http://www.innovations-report.com/html/reports/physics_astronomy/research_sheds_light_early_star_formation_135941.html .

59. M. J. Turk, et al "The Formation of Population III Binaries from Cosmological Initial Conditions", Turk et al, .Science 31 July 2009: 601-605 DOI: 10.1126/science.1173540.

60. G. Cresci, et al, 2004, "Stellar Clusters Forming in the Blue Dwarf Galaxy NGC 5253", astro-ph/0411486 v2.

61. A. S. Bishop, *Project Sherwood — The U.S. Program in Controlled Fusion,* (Addison-Wesley Publishing Company, Inc., Reading, Mass.,1958) p 177-178.

62. R. Clay and B. Dawson, Cosmic Bullets — *High Energy Particles in Astrophysics,* (Helix Books, Addison-Wesley, Australia, 1998).

63. Science Daily (Aug. 6, 2009) "Dark Energy From The Ground Up:Make Way For BigBOSS", http://www.sciencedaily.com/releases/2009/08/0908070910 28.htm .

64. D.J. Schlegal et al, 2009, "BigBOSS: The Ground-Based Stage IV Dark Energy Experiment", arXiv: 0904.0468 v3.

65. RIA Novosti (19/08/2009) "Russia, Germany to jointly probe for 'dark matter' [via soft x-ray emission]", http://en.rian.ru/science/20090819/155859444.html .

66. Space Daily (Aug. 19, 2009) "DLR And Roskosmos Sign Agreement On eRosita X-ray Telescope", http://www.spacedaily.com/reports/DLR_And_Roskosmos _Sign_Agreement_On_eROSITA_X_ray_Telescope_999.h tml .

67. V.C.Rubin, N. Thonnard and W. K. Ford, 1978 *Astrophys. J.* (Lett) 225, L107.

68. V. Rubin, *Bright Galaxies – Dark Matters,* (Amer. Inst. Physics, New York, 1997), p. 109-116.

69. S. Hawking, Lecture – "Life in the Universe", http://www.hawking.org.uk/lectures/life.html

70. S. Koppes, The University of Chicago Chronicle, Vol. 24, No.5, November 18, 2004, http://chronicle.uchicago.edu/041118/entropy.shtml

71. M. J. Jee, et al, 2007, arXiv: 0705.2171, "Discovery of a Ringlike Dark Matter Structure in the Core of the Galaxy Cluster Cl 0024+17".

72. Pierre Auger Collaboration, "Correlation of the Highest-Energy Cosmic Rays with Nearby Extragalactic..." Science 9 Nov. 2007, Vol.318, no. 5852, p. 938-943.

73. E. Pierpaoli and G. Farrar, 2005, "Massive galaxy clusters and the origin of Ultra High Energy Cosmic Rays" astro-ph/0 507679 v3.

74. P. Ricon, BBC, "Hubble makes 3D dark matter map" http://news.bbc.co.uk/2/hi/science/nature/6235751.stm .

75. S. Coleman, S. L. Glashow, 1998, "Evading the GZK Cosmic-Ray Cutoff," arXiv:hep-ph/9808446.

76. Discover magazine (Oct. 25, 2005) "Lorentz invariance" http://blogs.discovermagazine.com/cosmicvariance/2005/1 0/25/lorentz-invariance-and-you/ .

77. C.A. Scharf, et al, 2004, astro- ph/0406216.

78. M. S. Turner, 2001, arXiv:astro-ph/0108103, "Dark Energy and the New Cosmology".

79. R. Aloisio & V.S. Berezinsky, 2005, astro-ph/0507325

80. J.N. Bahcall and J.P. Ostriker, *Unsolved Problems in Astrophysics,* (Princeton Univ. Press, Princeton New Jersey, 1997), Chapter 17, The Highest Energy Cosmic Rays.

BIBLIOGRAPHY
AND SUGGESTED SOURCES

J. N. Bahcall and J. P. Ostriker, *Unsolved Problems in Astrophysics* (Princeton Univ. Press, Princeton, New Jersey, 1997).

A. S. Bishop, *Project Sherwood – The U.S. Program in Controlled Fusion* (Addison-Wesley Publishing Company, Inc., Reading, Massachusetts, U.S.A. 1958).

R. Clay and B. Dawson, *Cosmic Bullets – High Energy Particles in Astrophysics* (Helix Books, Addison-Wesley, Australia, 1998).

K. Croswell, *The Alchemy of the Heavens* (Anchor Books, Doubleday, New York, 1995).

J. Drexler, *How Dark Matter Created Dark Energy And The Sun* (Universal Publishers, Parkland, Florida, USA, 2003).

J. Drexler, *Comprehending and Decoding the Cosmos* (Universal Publishers, Boca Raton, Florida, USA, 2006).

J. Drexler, *Discovering Postmodern Cosmology* (Universal Publishers, Boca Raton, Florida, USA, 2008).

D. Filkin, *Stephen Hawking's Universe* (Basic Books, New York, 1997).

M. W. Friedlander, *A Thin Cosmic Rain – Particles From Outer Space* (Harvard Univ. Press, Cambridge, MA and London, England, 2000).

H. Friedman, *The Astronomer's Universe* (W.W. Norton & Company, New York, 1998).

T.K. Gaisser, *Cosmic Rays and Particle Physics* (Cambridge University Press, Cambridge U.K., 1990).

A. H. Guth, *The Inflationary Universe* (Helix Books, Perseus Books, Reading, Massachusetts, 1998).

R. P. Kirshner, *The Extravagant Universe – Exploding Stars, Dark Energy and the Accelerating Cosmos* (Princeton Univ. Press, Princeton, New Jersey, 2002).

E. W. Kolb and M. S. Turner, *The Early Universe,* (Addison-Wesley, USA, 1990).

L. Krauss, *Quintessence* (Basic Books, A Member of the Perseus Books Group, New York, N.Y., 2000).

T. S. Kuhn, *The Structure of Scientific Revolutions* (The University of Chicago Press, Chicago Illinois, 1970).

M. S. Longair, *High Energy Astrophysics,* Volume I, Second Edition (Cambridge Univ. Press, Cambridge, UK 1992).

M. S. Longair, *High Energy Astrophysics*, Volume II, Second Edition (Cambridge Univ. Press, Cambridge, UK, 1994).

M. S. Madsen, *The Dynamic Cosmos* (Chapman & Hall, London, England, 1995).

V. Rubin, *Bright Galaxies – Dark Matters* (Amer. Inst. Physics, New York, 1997).

G. Rudiger and R. Hollerbach, *The Magnetic Universe – Geophysical and Astrophysical Dynamo Theory* (Wiley-

VCH Verlag GmbH & Co. KGaA, Weinheim, Germany, 2004).

D. W. Sciama, *Modern Cosmology and the Dark Matter Problem* (Cambridge Univ. Press, Cambridge, UK, 1993).

P. Sokolsky, *Introduction to Ultrahigh Energy Cosmic Ray Physics* (Westview Press, A member of the Perseus Books Group, Boulder Colorado, USA, 2004).

T. Stanev, *High Energy Cosmic Rays* (Springer-Verlag Berlin Heidelberg New York, 2004)

T. X. Trinh, *The Secret Melody* (Oxford Univ. Press, New York, 1995).

W. S. C. Williams, *Nuclear and Particle Physics* (Oxford Univ. Press, Oxford and New York, 1991).

S. Yoshida, *Ultra-High Energy Particle Astrophysics* (Nova Science Publishers, Inc., New York N.Y., 2003).

GLOSSARY

Accelerating Cosmos: Accelerating expansion of universe

Accretion: An infall of matter onto an object.

AGASA: Akeno Giant Air Shower Array.

AGN: Galaxies having active galactic nuclei.

Alpha Particle: The nucleus of a helium atom.

Andromeda Galaxy: Twin galaxy of the Milky Way. The two galaxies comprise most of the Local Group's mass.

Astro-Cosmology: Astronomy oriented dark matter cosmology

Astronomical Unit (A.U.): The average distance from the Sun to the Earth, equal to 149,598,000 kilometers.

Astrophysics: The study of the composition and other physical properties of celestial objects.

Astrophysical Emergence: See emergence and emergent evolution.

Astrophysical Dynamo Effect: Relativistic protons orbiting galaxies will create magnetic fields through the astrophysical dynamo effect under which the relativistic protons moving in Larmor orbits create magnetic fields. These same magnetic fields in turn determine the proton paths, eventually reaching a steady-state solution for the magnetic fields and the proton paths after an emergent evolution period involving millions to billions of years.

Baryon/Baryonic: An elementary particle that is subject to the strong nuclear interaction. The proton and neutron and combinations of them are baryons.

Big Bang: The cosmological theory that all the matter and energy in the Universe was concentrated in an immensely hot and dense point, which exploded 13.7 billion years ago.

BigBOSS: American planned ground-based high-precision spectrograph to be use to research dark energy and the post-big bang rapid expansion of the universe.

Black Hole: An object that exerts such enormous gravitational force that nothing, not even light or other forms of electromagnetic radiation, can escape from it.

BL Lac, BL Lacertae Objects: Objects that are in the category of AGN's, now known as blazars, that exhibit no emission lines but have a continuum emission from radio frequencies through X-ray frequencies.

Bottom-Up Theory: Theory that small galaxies form first and larger galaxies are formed through mergers of small galaxies.

Bremsstrahlung Radiation: Braking radiation of a proton or other charged particle.

CERN: Center for European Nuclear Research.

Chandra X-ray Observatory: Part of NASA's fleet of "Great Observatories" along with the Hubble Space Telescope. Chandra allows scientists to obtain unprecedented X-ray images of exotic environments.

Closed Universe: A universe in which the density of matter is greater than the critical density and that should thus collapse onto itself in the future.

Cluster Soft Excess (CSE): EUV and soft X-ray emission from a galaxy cluster beyond what would be expected based upon its temperature.

COBE: Cosmic Background Explorer.

Cold Dark Matter (CDM): Non-baryonic matter consisting of putative elementary particles of relatively high mass that are moving relatively slowly. Invented in 1984.

Collective Self-organization: A term very similar in meaning to emergence and emergent evolution.

Collision Cross Section: A measure of the probability that an encounter between particles will result in the occurrence of a particular atomic or nuclear reaction.

Coma Cluster: A galaxy cluster that contains about 1,000 galaxies. The gravitational effects of dark matter were discovered in this galaxy cluster.

Comet: A body of ice and dust, with a nucleus of typically about 10 kilometers in diameter.

Copernicus' (Nicolaus) Concept: Earth rotates daily and the planets revolve in orbits around the Sun. He authored *On The Revolutions of Heavenly Spheres* published in1543.

Cosmic DM Mysteries/Cosmic Constituents: Dark matter mysteries or unexplained phenomena regarding celestial bodies or cosmic matter such as their shape, mass distribution, particle abundance ratios, dimensions, density, location, maturity, acceleration, velocity, linear momentum, angular momentum, particle energies, star rotation curves, hydrogen fusion reactions, particle energy distributions, particle transformations, star ignition, and star formation rates.

Cosmic Inflation: The hyper rapid expansion of the Universe a fraction of a second after the Big Bang. See Chapters 21 and H.

Cosmic Microwave Background (CMB) Or Cosmic Background Radiation (CBR): The microwave radiation that bathes the entire Universe and that dates from the epoch when the Universe was just 300,000 years old.

Cosmic Origins Spectrograph (COS): American ultraviolet space spectrograph installed on the Hubble Space Telescope May 2009.

Cosmic-ray Cosmology: A new term to describe a cosmology recently developed by J. Drexler, based upon UHE protons and cosmic ray protons.

Cosmic Rays: Particles (mostly protons and electrons) that have been accelerated somewhere in the Universe to very high energies.

Cosmic Web: The cosmic web is considered to be the framework on which the universe is built. It is comprised primarily of dark matter that makes up about 83 percent of the mass of the universe. It is explained in Chapter 33.

Cosmology: The study of the Universe as a whole, and of its structure and evolution.

Cosmos: An orderly, harmonious, and systematic Universe.

Coulomb Force: The force between two coulomb charges or electrically charged particles.

CSE: See Cluster Soft Excess.

Dark Energy (as defined in the past): A hypothetical form of energy that permeates all space and has negative pressure resulting in a repulsive gravitational force. The accelerating expansion of the Universe has been attributed to dark energy. Beginning 2008, see Chapter 21.

Dark Galaxy: A galaxy with a total light output or luminous level from stars below an established minimum threshold level. Such galaxies represent extreme cases of low surface brightness (LSB) galaxies.

Dark Matter (as defined in the past): Matter that is detected only by its gravitational pull on visible matter. The composition has been unknown; it might consist of very low mass stars or supermassive black holes, but Big Bang nucleosynthesis calculations limit the amount of such baryonic matter to a small fraction of the critical mass density. If the mass density is critical, as predicted by the simplest versions of inflation, then the bulk of the dark matter must be a gas of weakly interacting non-baryonic particles, sometimes called WIMPs (Weakly Interacting Massive Particles).

Deuterium: A chemical element whose nucleus consists of a proton and a neutron, created mainly in the first three minutes of the Universe's history.

DOE: U.S. Department of Energy.

Doppler Effect: The variation in the energy and color of light caused by the motion of a source of light relative to an observer. If the source is receding, the energy decreases and the light is shifted toward the red. If the source is approaching, the energy increases and the light is shifted toward the blue.

Doppler Shift: The shift in the received frequency and wavelength of an electromagnetic wave that occurs when either the source or the observer is in motion. Approach causes a shift toward shorter wavelengths and higher frequencies called a blue shift. Recession has the opposite effect, called a red shift. The expansion of the Universe causes ancient electromagnetic wave emissions to exhibit a doppler red shift.

Drexler Dark Matter: See Relativistic-Proton Dark Matter.

Dwarf Galaxy: A galaxy with a small size and mass. The average diameter is about 15,000 light-years; that is, about one-sixth of that of a normal galaxy. Masses range from 100 million to 1 billion solar masses, about 1,000 to 10,000 times less than the mass of an ordinary galaxy. Dwarf galaxies may be spheroidal or irregular, but dwarf spiral galaxies have not been observed.

Dwarf Spheroidal Galaxy (dSph): A dwarf galaxy that is spheroidal in shape, has an old stellar population, and lies close to a large host galaxy as a satellite.

Dwarf Irregular Galaxy (dIrr): A dwarf galaxy that is irregular in shape, has a young stellar population, and is a satellite of a large host galaxy, but lies further away from its host galaxy than would a dwarf spheroidal galaxy.

EeV: 10^{18} eV.

Electromagnetic Wave: A pattern of electric and magnetic fields that moves through space. Depending on the wavelength, an electromagnetic wave can be a radio wave, a microwave, an infrared wave, a wave of visible light, an ultraviolet wave, a beam of X rays, or a beam of gamma rays.

Electron: The lightest of the subatomic particles with electrical charge. The electron has a mass of 9×10^{-28} kilograms and is negatively charged.

Electron Volt (eV): The energy released when a single electron passes through a one-volt battery.

Elliptical Galaxy: A galaxy observed as an oval-shaped system generally composed of old stars, a large black hole, and containing little or no gas and dust.

Emergence, Emergent Evolution: A theory that new characteristics and qualities appear in the evolutionary process at more complex organizational levels (than that of the pre-existent entities such as a molecule, a cell, or a particle) and which cannot be predicted solely by studying less complex levels of organization but which are determined by a rearrangement of pre-existent entities.

Entropy: A measure of the unavailability of a system's energy to do work.

eROSITA: German X-ray Space Telescope

ESO: European Space Organization.

EUV: Extreme ultraviolet radiation.

Extragalactic, Intergalactic: The regions of the Universe outside of any galaxy.

Field Galaxies: The variety of galaxy types typically found in galaxy surveys.

FUV: Far-ultraviolet spectrum

Galactic Disk: A flattened aggregation of stars, gas, and dust in a spiral galaxy. The average disk is some 90,000 light-years in diameter and 300 light-years thick. In the Milky Way, the stars complete one turn around the galactic center every 250 million years, at a velocity of 230 kilometers per second.

Galactic Halo: A spherical region around a spiral galaxy populated by old stars and globular clusters. Observations suggest that it is surrounded by a dark matter halo some 10 to 20 times larger than the galaxy and more massive.

Galaxy: A system of stars (10 million in a dwarf galaxy, 100-200 billion in an average galaxy like the Milky Way, 10 trillion in a giant galaxy) held together by gravity.

Galaxy Cluster: A dense grouping of several thousand galaxies bound by gravity, with an average diameter of some 60 million light-years, and an average mass of a few million billion solar masses.

GALEX, Galaxy Evolution Explorer: American ultraviolet space telescope

Gamma Ray: An electromagnetic wave with a wavelength in the range of 10^{-13} to 10^{-10} meters, corresponding to photons with energy in the range of 10^4 to 10^7 electron volts. Their energies are higher than X rays.

Gauss: A measure of the strength of a magnetic field.

General Relativity: A gravitational theory proposed by Albert Einstein in 1915, which is more accurate than that of Newton. The two theories differ mainly in situations where gravitational fields are very intense.general relativity supports the big bang theory.

GeV: G stands for Giga, or 10^9. Thus, GeV is one billion electron volts.

Gravitational Field: A field of force surrounding a body of finite mass. The field of force is defined as the force that would be experienced by a standard mass positioned at each point in the field.

Gravitational Tidal Force: The tidal force responsible for attraction between all matter. The weakest of the four forces, gravitational force possesses the longest range.

GRB: A gamma ray burst.

Group of Galaxies: A collection of about 20 galaxies held together by gravity, some six million light-years across and averaging between one and 10 trillion solar masses.

Gyr: Gigayear, or one billion years.

GZK Cosmic-Ray Cutoff: A theory limiting proton energies. According to the currently questioned 1966 Greisen-Zatsepin-Kuzmin (GZK) cutoff theory, protons with energies greater than $6x10^{19}$ eV would interact with the cosmic microwave background radiation and lose energy through radiation and thus would not travel more than 50 Mpc, or about 160 million light-years. In 1998, Coleman and Glashow wrote a paper entitled, "Evading the GZK Cosmic-Ray Cutoff," which showed that for very high energy cosmic rays, the GZK cutoff may not apply.

Halo: The region around a galaxy that contains dark matter and some halo stars.

Helium: A chemical element with a nucleus of two protons and two neutrons (helium-4). A second, far-less-abundant isotope has two protons and one neutron (helium-3).

HESS High Energy Spectrographic System: A system used for detecting gamma rays.

Hubble Law: The law discovered in 1929 by the American astronomer Edwin Hubble, which states that the distance of galaxies varies in proportion to their red shift and, thus, because of the Doppler effect, to their velocity of recession. The law gave birth to the idea of an expanding universe.

Hydrogen: The lightest of all chemical elements, consisting of one proton and one electron. Hydrogen makes up 75% of the mass of the Universe.

Isotropy/Isotropic: The property of the Universe to be similar in every direction.

Kpc: The abbreviation for a kilo parsec where a parsec equals 3.26 light-years.

Kepler's laws: Laws concerning the motions of the planets in their orbits derived by Johannes Kepler in the 16th century.

Large Magellanic Cloud (LMC): The larger of two irregularly shaped galaxies closest to the Milky Way located in the far southern sky and visible to the unaided eye.

Larmor Radius (for a proton): A proton crossing an orthogonal/magnetic field and entering into a spiral path. The radius of a cycle of that spiral path is called the proton Larmor Radius for that cycle.

$$\text{Proton Larmor Radius} = 110 \text{ Kpc} \times \frac{10^{-8} \text{ gauss}}{B} \times \frac{E}{10^{18} \text{ eV}}$$

Law of Conservation of Linear Momentum: The total linear momentum of the mass objects in a group remains unchanged. See momentum.

Light-year: The distance traveled by light (which moves at a velocity of 300,000 kilometers per second) in one year and equal to 9,460 billion kilometers.

Local Group: A grouping of galaxies extending over a region of space of about 10 million light-years, of which the Milky Way and Andromeda are the principal and most massive members (one trillion solar masses each). It also includes dwarf galaxies.

Lorentz Invariance: A relativistic proton crossing magnetic field lines would violate Lorentz invariance because a magnetic field is vector field, not a scalar field.

Low-Surface Brightness (LSB) Galaxy: A diffuse galaxy with a surface brightness that is one magnitude lower than the ambient night sky.

Lyman-Alpha Blob: Huge concentration of protons, electrons, and hydrogen gas in the early universe emitting the Lyman-alpha emission line.

Lyman-Alpha Emission Line: Emission of ultraviolet photons at a wavelength of 121.6 nanometers when a proton and electron combine to form hydrogen.

Magnetic Bulge: A significant rise in the orthogonal magnetic field experienced by a relativistic proton.

Magnetic Field: A field of force in space, created by a magnet or by an electric current, that guides the trajectories of electrically charged particles by exerting an electromagnetic force.

Mass: The measure of the inertia of an object, determined by observing the acceleration when a known force is applied. An object with mass creates a gravitational field, which is defined in this glossary. When a proton travels at relativistic velocities, it has a relativistic mass equal to its energy divided by the square of the speed of light.

Maxwell's Equations: A set of differential equations describing space and time dependence of the electromagnetic field and forming the basis for classical electrodynamics.

Microwave: An electromagnetic wave with a wavelength of between one millimeter and 30 centimeters.

Milky Way: The galaxy to which our solar system belongs, whose central regions appear as a band of light or "milky way" that we can see from Earth in clear night skies.

Missing Baryons: About 4.5 percent of the mass/energy of the universe is estimated to be baryonic mass. It can be confirmed as of 10 billion years ago, but half seems to be missing now. .

Missing Mass: An outmoded name for the dark matter of the Universe.

MNRAS: Monthly Notes of the Royal Astronomical Society.

m_0: The symbol m_0 representing the mass of a proton when it is not moving (the rest mass).

Momentum: The linear momentum of an object, equaling the product of its mass and velocity. If no external forces are acting on a group of mass objects, the Law of Conservation of Linear Momentum requires that the total linear momentum of the mass objects in the group remains unchanged.

MOND, Modified Newtonion Dynamics: Theory of gravity challenging gravity-based dark matter theories.

Muon (contraction of the earlier mu-meson; taken as a symbol for meson, and used to distinguish it from the short-lived pi-meson): An unstable elementary particle that belongs to the lepton family, that is common in the cosmic radiation near the Earth's surface, that has a mass about 207 times the mass of the electron, and that exists in negative and positive forms.

Muonic Ion: Two nuclei of atoms in close proximity, usually one of them being a proton and the other being a deuteron, a helium nucleus, or another proton, being orbited very closely by a single negative muon weighing 207 times as much as an electron at rest. Muonic ions are best known for catalyzing low temperature nuclear fusion reactions.

NASA: National Aeronautics and Space Administration.

Neutralino: A theoretical non-baryonic particle, which is an amalgam of the superpartners of the photon (which transmits the electromagnetic force), the Z boson (which transmits the so-called weak nuclear force), and perhaps other particle types. Although the neutralino is heavy by normal standards (at least 35 times the mass of a proton), it is generally thought to be the lightest supersymmetric particle.

Neutron: A subatomic particle with no electric charge, one of the two basic constituents of an atomic nucleus.

Nucleosythesis: The production of a chemical element from hydrogen nuclei or protons, as in stellar evolution.

Ockham's Razor Logic (also called Occam's razor): A scientific and philosophic rule that the favored explanation for an unknown phenomenon is the simplest of the competing theories. It should be preferred to the more complex, or that explanations of unknown phenomena be sought first in terms of known quantities rather than through assumptions.

NSF: National Science Foundation.

Open Universe: A Universe in which the density of matter is less than the critical density and which will thus expand forever.

Oort Cloud: A region in the outer limits of the solar system where billions to trillions of comets reside.

Orthogonal: Intersecting or lying at right angles.

Parsec: An astronomical unit of distance equal to 3.26 light-years or approximately 19 trillion miles.

Pion (contraction of pi-meson): A short-lived meson that is primarily responsible for the nuclear force and that exists as a positive or negative particle with mass 273.2 times the electron mass or a neutral particle with mass 264.2 times the electron mass.

Population I Stars: A younger generation of stars with ages from a few million years to about 10 billion years and with a relatively large fractional abundance (about 1% of mass) of elements heavier than helium. The Sun is in this category.

Postmodern Cosmology or Postmodern Big Bang Cosmology: A recently discovered cosmology in which the output of the Big Bang is comprised almost entirely of relativistic protons and helium nuclei in a ratio of about 12:1 and dark matter is comprised of the identical subatomic particles in the essentially the same ratio.

Primordial Ripples: The mass perturbations in the early Universe that may have evolved into galaxies.

Proto-galaxy: A cloud of gas and ions that is evolving into a galaxy.

Proton: A positively charged particle composed of three quarks that, together with the neutron, forms atomic nuclei. The proton is 1,836 times more massive than the electron.

RBB: Relativistic Big Bang. See Chapters 12, 14, and 35.

Recombination: In the traditional theory, between 300,000 and 700,000 years after the Big Bang, the plasma of free electrons and hydrogen nuclei that condensed to form a neutral gas, in a process called recombination. The prefix "re" is not meaningful here, however, since according to the Big Bang theory, the electrons and protons (hydrogen nuclei) were combining for the first time ever.

Red Shift: A shift to longer wavelengths and lower frequencies, typically caused by the Doppler effect in a receding object or caused by the expansion of the Universe.

Reductionism: A procedure or theory that reduces or attempts to reduce complex data or phenomena to simple elements or terms. It is an inward-looking approach that in physics usually means a search for subatomic particles in an attempt to understand or explain some unusual phenomenon.

Relationism: An analytical procedure, method, concept, or theory, developed by Jerome Drexler, that attempts to identify dark matter by determining which cosmic phenomena may be facilitated, expedited, influenced by, or have a special relationship with dark matter. This outward-looking cosmological concept is used to determine the nature and characteristics of dark matter's influence on and relationship with cosmic phenomena as a means of DM identification.

Relativistic-Proton Dark Matter, Relativistic-Baryon Dark Matter, or **Drexler dark matter:** The ratio of relativistic protons to relativistic helium nuclei ranges between 10:1 and 12: 1. The dark matter particle streams form spheroidal halos around individual galaxies, orbit groups of galaxies within galaxy clusters, and form large long filaments of dark matter that comprise the Cosmic Web.

Rotation Curve of a Galaxy: A graph of the orbital velocities of stars or hydrogen as a function of their radial distances from the nucleus of the galaxy radially outward into the surrounding dark matter halo.

R-PDM: Relativistic-Proton dark matter.

Schmidt Law: An empirical law, for isolated spiral galaxies, of the correlation between the star formation rate (SFR) and the overall average molecular hydrogen surface density.

Second Law of Thermodynamics: The entropy (disorder) of the Universe always increases over time.

SFR: Star formation rate.

Signature Characteristics (SigChar): An extensive list, for each dark matter candidate, of all the possible features or characteristics of the candidate that can be attributed to it by utilizing any and all laws and principles of physics.

Solar System: The Sun and the objects in orbit around it, which include nine planets, nearly 60 known satellites of the planets, thousands of smaller objects called asteroids, and billions to trillions of comets.

Spectrum-Roentgen-Gamma (SRG): Russian satellite to carry eROSITA German X-ray space telescope and planned Russian X-ray spectrograph in 2012.

Spektr-UF (Spectrum-UV): Russian ultraviolet space telescope planned for 2010.

Spiral Galaxy: A flattened, disk-like system of stars and interstellar gas and dust with a spherical collection of stars, known as the bulge, at its center. Bright, young stars outline spiral arms in the plane of the disk.

(The) Standard Model: Name given to the current theory of fundamental particles and how they interact.

Star: A sphere of gas consisting of 98% hydrogen and helium and 2% heavy elements in equilibrium under the action of two opposing forces — the compressive gravity and the outward radiation pressure from the nuclear fusion reactions in its core. The Sun has a mass of 2×10^{30} kilograms, and masses of stars range between 0.1 and 100 solar masses.

Starburst Galaxy: A galaxy experiencing a period of intense star forming activity. They are usually associated with the merging or interaction of two galaxies. This activity may last for 10 million years or more. During a starburst, stars can form at tens, even hundreds, of times greater rates than the star formation rate in normal spiral galaxies.

Subatomic Particles: They could be fast protons or helium nuclei or any fast atom with its electrons removed.

Supercluster: The aggregation of tens of thousands of galaxies held together by gravity and gathered into groups and clusters. Superclusters have the shape of flattened pancakes with an average diameter of 90 million light-years and masses of 10,000 trillion (10^{16}) solar masses.

Supernova/Supernovae: An exploding star, visible for weeks or months, even at enormous distances, because of the tremendous amounts of energy that the star produces. Supernovae typically arise when massive stars exhaust all means of producing energy from nuclear fusion. In these stars, the collapse of the star's core results in the explosion of the star's outer layers. Another type of supernova arises when hydrogen-rich matter from a companion star accumulates on the surface of a white dwarf and then undergoes nuclear fusion. This second type, known as a Type 1a supernova, generates light at a well-known standard level and thus can be used to measure the rate of expansion of the Universe.

Synchrotron Emission or Radiation: Electromagnetic radiation that is emitted by charged particles moving at relativistic speeds in circular orbits in a magnetic field. The rate of emission is inversely proportional to the product of the radius of curvature of the orbit and the fourth power of the mass of the particles. For this reason, synchrotron radiation is not a problem in the design of proton synchrotrons, but it is significant in electron synchrotrons. The synchrotron emission from a relativistic proton or electron is directly proportional to the square of its energy.

TeV: T stands for Tera, or 10^{12}. Thus, TeV is one trillion electron volts.

Tokamak: Tokamak Hydrogen Fusion Test Reactor

Top-Down Theory: The theory that long, large, dark matter filaments of the cosmic web form galaxy clusters where the DM filaments intersect/collide and then galaxies form from the remnants of these collisions.

UHE (Ultra-High Energy) Proton: A proton traveling near the speed of light with an energy of at least 10^{18} eV.

UHECR: Ultra-high energy cosmic ray proton traveling near the speed of light with an energy of at least 10^{18} eV.

Ultraviolet (UV): Ultraviolet light.

Virgo Supercluster: A huge, flattened supercluster that contains the Local Group of galaxies. The Local Group, containing the Milky Way, lies at the edge of the supercluster, while the Virgo Cluster of galaxies is at its center.

White Dwarf Star: A small, dense star with a diameter of about 10,000 kilometers (about the size of Earth) created when a star of less than 1.4 solar masses exhausts the nuclear fuel and collapses under its own gravity. This type of star participates in a Type 1a supernova.

WIMP (Weakly Interacting Massive Particle): The name for a non-baryonic theoretical dark matter candidate that is presumed to have a mass much greater than that of a proton. The theoretical neutralino is one form of WIMP.

WMAP (Wilkinson Microwave Anisotropy Probe): A NASA Explorer mission measuring the temperature of the cosmic background radiation over the full sky. This map of the remnant heat of the Big Bang provides data about the origin of the Universe.

X Rays: Electromagnetic radiation with greater frequencies and smaller wavelengths than those of ultraviolet radiation and lower frequencies and longer wavelengths than those of gamma ray radiation.

INDEX

Our Universe
via Drexler Dark Matter

Synopsis

This book is different from all other modern cosmology books in several ways. It introduces a cosmologic universe, which is orderly, logical, and systematic. It teaches and explains by illustrating how a variety of cosmic mysteries have been solved. It raises the status of dark matter in the universe by illuminating its roles as the principal source of energy, the principal source of matter in the form of hydrogen and helium, and the principal source of cosmic relationships with the principal cosmic phenomena of the universe. This book simplifies the universe as Nicolaus Copernicus' book simplified the solar system in 1543.

With more and more cosmic mysteries being discovered and the slow progress in solving them, cosmologists and astrophysicists must re-train themselves to understand and to utilize the postmodern unified astrophysical cosmology model and to maximize the knowledge derived from the astronomical data. These are the three principal objectives of this book.

About the Author

Jerome Drexler is a former member of the technical staff and group supervisor at Bell Labs, former research professor in physics at New Jersey Institute of Technology, founder and former Chairman and chief scientist of LaserCard Corp. (Nasdaq: LCRD). He has been awarded 76 U.S. patents, honorary Doctor of Science degrees from NJIT and Upsala College, a degree of Honorary Fellow of the Technion, an Alfred P. Sloan Fellowship at Stanford University, a three-year Bell Labs graduate study fellowship, the 1990 "Inventor of the Year Award" for Silicon Valley and recognition as the original inventor in 1978 of the now widely-used digital optical disk "Laser Optical Storage System" and the LaserCard(R) nanotech data memory. He is a member of the Board of Overseers of New Jersey Institute of Technology and an Honorary Life Member of the Technion Board of Governors.